Emerging Technologies in Hydraulic Fracturing and Gas Flow Modelling

*Edited by Kenneth Imo-Imo Israel Eshiet
and Rouzbeh G. Moghanloo*

Published in London, United Kingdom

IntechOpen

Supporting open minds since 2005

Emerging Technologies in Hydraulic Fracturing and Gas Flow Modelling
http://dx.doi.org/10.5772/intechopen.87833
Edited by Kenneth Imo-Imo Israel Eshiet and Rouzbeh G. Moghanloo

Contributors
Yaxiong Li, Zhiming Hu, Yanran Li, Hisham Ben Mahmud, Ziad Bennour, Walid Mohamed Mahmud, Mansur Ermila, Javed Akbar Khan, Eswaran Padmanabhan, Izhar Ul Haq, Duvvuri Satya Subrahmanyam, Mileva Radonjic, Gabriel Adua Awejori, Kenneth Imo-Imo Israel Eshiet

Notice
Statements and opinions expressed in the chapters are these of the individual contributors and not necessarily those of the editors or publisher. No responsibility is accepted for the accuracy of information contained in the published chapters. The publisher assumes no responsibility for any damage or injury to persons or property arising out of the use of any materials, instructions, methods or ideas contained in the book.

First published in London, United Kingdom, 2022 by IntechOpen
IntechOpen is the global imprint of INTECHOPEN LIMITED, registered in England and Wales, registration number: 11086078, 5 Princes Gate Court, London, SW7 2QJ, United Kingdom
Printed in Croatia

British Library Cataloguing-in-Publication Data
A catalogue record for this book is available from the British Library

Additional hard and PDF copies can be obtained from orders@intechopen.com

Emerging Technologies in Hydraulic Fracturing and Gas Flow Modelling
Edited by Kenneth Imo-Imo Israel Eshiet and Rouzbeh G. Moghanloo
p. cm.
Print ISBN 978-1-83968-466-1
Online ISBN 978-1-83968-467-8
eBook (PDF) ISBN 978-1-83968-468-5

We are IntechOpen,
the world's leading publisher of
Open Access books
Built by scientists, for scientists

6,100+
Open access books available

149,000+
International authors and editors

185M+
Downloads

Our authors are among the

156
Countries delivered to

Top 1%
most cited scientists

12.2%
Contributors from top 500 universities

Interested in publishing with us?
Contact book.department@intechopen.com

Numbers displayed above are based on latest data collected.
For more information visit www.intechopen.com

Meet the editors

Kenneth Imo-Imo Israel Eshiet is a senior lecturer at the University of Wolverhampton, United Kingdom. Prior to his current position, he was on different occasions an assistant professor at Prince Mohammad Bin Fahd University, Saudi Arabia, a researcher at the University of Leeds, United Kingdom, a lecturer at the University of Uyo, Nigeria, and a senior consultant at Sustainable Energy Environmental and Educational Development (SEEED), USA. He is a chartered civil engineer in the United Kingdom with professional interests in areas including the development of numerical/analytical methods for engineering problems; experimental and numerical modelling of structures and geotechnical systems; site investigation and laboratory and field geotechnical experimentation; computational fluid dynamics; stochastic and optimisation analysis; and structural analysis and design.

Dr. Rouzbeh G. Moghanloo is an associate professor at the Petroleum and Geological Engineering Department, University of Oklahoma, USA. Dr. Moghanloo received his BS and MS in Chemical Engineering from the Amirkabir University of Technology, Iran, and obtained a Ph.D. in Petroleum Engineering from the University of Texas at Austin, USA. He has been an active member of numerous technical societies and served as an associate editor for multiple technical journals. Dr. Moghanloo has worked as a reservoir engineer and technical advisor for numerous companies. He is a recipient of the 2018 SPE Mid-Continent Regional Reservoir Description and Dynamics Award and the 2016 American Chemical Society award.

Contents

Preface   XI

Section 1
Hydraulic Fracturing for Oil/Gas Recovery   1

Chapter 1   3
Production from Unconventional Petroleum Reservoirs:
Précis of Stimulation Techniques and Fluid Systems
by Kenneth Imo-Imo Israel Eshiet

Chapter 2   33
A Review of Fracturing Technologies Utilized in Shale Gas Resources
by Hisham Ben Mahmud, Mansur Ermila, Ziad Bennour
and Walid Mohamed Mahmud

Chapter 3   57
Hydraulic Fracturing in Porous and Fractured Rocks
by Duvvuri Satya Subrahmanyam

Chapter 4   93
Hydraulic Fracture Conductivity in Shale Reservoirs
by Javed Akbar Khan, Eswaran Padmanabhan and Izhar Ul Haq

Chapter 5   113
Review of Geochemical and Geo-Mechanical Impact of Clay-Fluid
Interactions Relevant to Hydraulic Fracturing
by Gabriel Adua Awejori and Mileva Radonjic

Section 2
The Gas Flow Model   139

Chapter 6   141
Mechanism, Model, and Upscaling of the Gas Flow in Shale Matrix:
Revisit
by Zhiming Hu, Yaxiong Li and Yanran Li

Preface

Despite the clamor for reduced dependency on fossil fuels, the demand for hydrocarbon-based energy is on the rise. Consequently, the strain on oil and gas reservoirs is greater than ever before. This has also drawn more attention to unconventional reservoirs, making it exigent for further innovations in producing hydrocarbon formations. This book introduces an array of cutting-edge strategies for the exploitation of depleted and unconventional reservoirs. It provides some insights pertaining to interactions between formation in situ and injected fluids and rock constituents under extreme underground conditions. In addition, it presents a novel model for gas flow, pertinent to understanding flow mechanisms in the shale rock matrix.

In tandem with the evolution of the petroleum industry, there have been major advancements in the range of stimulation methods for the production of reservoirs. Chapter 1 is a breakdown of the different types of hard-to-produce rock formations and the corresponding diversity of stimulation techniques appropriate for their production. Furthermore, the chapter describes the multiplicity of fracturing fluid systems and the associated influencing factors, which are instrumental to the successful execution of many stimulation operations. This is then focalised on shale reservoirs in Chapter 2, which casts a spotlight on the fracturing mechanisms and evolvement in fracturing technologies applied in shale gas formations. In Chapter 3, a case is presented that supports the extended application of hydraulic fracturing techniques in the measurement of in situ stresses in porous and fractured rocks. This approach promises better results at deep subsurface levels in comparison to antecedent techniques. Chapter 4 demonstrates formation-proppant contact behavior and its impact on the conductivity of shale reservoirs following the creation of hydraulic fracture networks with the aid of a model that analyzes contact characteristics of proppant embedment. Chapter 5 is a detailed review of the interactions between formation/hydraulic fracturing fluids and clay minerals in shale formations, with an emphasis on geomechanical and geochemical feedback. Chapter 6 presents a nanoscale model to investigate the gas flow mechanisms of shale reservoirs. The chapter also explores upscaling techniques used to translate results from the proposed model and laboratory-scale experiments to field-scale representations.

The state-of-the-art techniques and procedures exhibited in this book are yet another indication of the pool of resounding efforts to harness unconventional hydrocarbon resources more effectively. Applications of the proposed methods are not limited to the scenarios presented; it is possible to extend them to a broader ambit of underground conditions, provided the constraints are recognised.

Dr. Kenneth Imo-Imo Israel Eshiet
Faculty of Science Engineering,
University of Wolverhampton,
Wolverhampton, United Kingdom

Rouzbeh G. Moghanloo
University of Oklahoma,
Oklahoma, United States of America

Section 1

Hydraulic Fracturing for Oil/Gas Recovery

Production from Unconventional Petroleum Reservoirs: Précis of Stimulation Techniques and Fluid Systems

Kenneth Imo-Imo Israel Eshiet

Abstract

An overview of the different categories of unconventional oil and gas reservoirs, and corresponding stimulation techniques appropriate for them is examined. Three main groups of unconventional oil and gas formations are appraised: heavy oil, oil shale and tight reservoirs. The scope of stimulation methods applicable to heavy oil reservoirs is limited. This kind of formation contains characteristic high-viscous hydrocarbons and are produced majorly by cold production and thermal stimulation. On the other hand, a wider range of stimulation methods are successfully used to produce tight and oil shales formations. For oil shales, these include drilling horizontal wells as substitutes to vertical wells, hydraulic fracturing, surfactant treatment, water imbibition, thermal treatment and acidisation; whilst for tight formations, these include hydraulic fracturing, surfactant treatment, water imbibition, acidisation and the application of electro-kinetics. Fracturing fluid systems are integral to the implementation of most stimulation operations and are evaluated herein under the following groups: water-based, oil-based, foam-based and acid-based. The most commonly used fracturing fluids are water based, albeit there are several instances where other types of fluids or combination of fluids are more suitable based on factors such as formation sensitivity, costs, wettability, rock solubility, surface tension, capillarity, viscosity, density, rheology and reactivity.

Keywords: unconventional reservoirs, reservoir stimulation, hydraulic fracturing, acidisation, surfactant, fracturing fluids, horizontal wells

1. Introduction

Unconventional hydrocarbon reservoirs are different from their conventional counterparts in the sense that they require distinctive operations for recovery that differ from normal practices deployed for conventional reservoirs. The main reason for this is the ultra-low permeability of the rock formation, which hinders the ease of flow of hydrocarbons towards the well, but other factors such as the reservoir fluid properties also impact flow mechanisms. Examples of unconventional reservoirs are gas hydrates, oil shales, gas shales, tight-gas sandstones, tight-gas limestones, heavy oil and tar sandstones, and coalbed methane reservoirs [1–7].

IntechOpen

As the term implies, heavy oil and tar reservoirs are those that contain viscous and dense oils. About a third of the total world oil and gas reserves consist of the heaviest range of hydrocarbons, yet they are mostly overlooked due to the perceived high costs and difficulties associated with its production [5]. Although reservoir properties including pressure, permeability and porosity are important measures of its behaviour, the fluid density and viscosity determine the approach used for production [5]. Heavy oils and tars are generally high in density and viscosity. Density is a measure of how much mass is contained per unit volume. The standard unit of measurement adopted in the oil and gas industry, especially in the United States, is the degree of American Petroleum Institute (API) gravity. A lower API value indicates a higher density and vice versa. Normally, oils below 20° API gravity are defined as heavy which may be as low as 4° for bitumen with high tar content [3, 5]. Oil viscosity, on the other hand, defines its resistance to gradual shear or tensile deformation when subjected to shear or tensile stress respectively. A viscous fluid exhibits resistance to shear stress and, thus, its flow is reduced where shear stresses are applied. Oil viscosity has an inverse relationship with temperature; it varies greatly by becoming less viscous as temperature increases. The flow rate of reservoir fluids is a key parameter and because of the direct link between viscosity, temperature and the ease of flow, oil viscosity is considered to be more important than oil density during production [3, 5]. Thus, viscosity, rather than density is used as a measure of the heaviness of oil. Under reservoir conditions, heavy oils have viscosities >100 cp [3]. Apparently, there is no direct correlation between density and viscosity, largely due to the influence of temperature. Low-density oils in shallow reservoirs, where the temperatures are cooler, may have higher viscosities in comparison to oils at hotter deep reservoirs.

Oil shale is a fine-grained sedimentary rock richly composed of organic matter [8], in the form of kerogen [2]. Kerogen is a solid mixture of organic compounds and is the primary source of hydrocarbons from oil shale. This type of hydrocarbon is referred to as shale oil, which is unconventional and different from tight oil naturally present in shales and ultra-low permeability sandstones, carbonates and siltstones [9]. Kerogen, also known in some instances as total or bitumen-free organic matter, consist of more than 80% organic matter; however, a major proportion of this is not readily soluble in ordinary organic solvents under moderate conditions [2]. Therefore, it is more challenging to extract in comparison to crude oil from conventional reservoirs because of high costs and negative environmental impacts [10]. To remove shale oil from oil shales, it is imperative to decompose the insoluble organic matter with heat. This is achieved by thermal dissolution, hydrogenation or pyrolysis [11–13]. The three methods require very high temperatures.

Tight oil is light crude oil found in shales and very low permeability and low porosity sandstones, carbonates and siltstones [9]. Although the term is sometimes used interchangeably with shale oil normally contained in oil shales (e.g., [9, 14]), there are distinctions. As at 2015 the world's technically recoverable tight oil from shale formations was estimated at 418.9 billion barrels (bbl). A large proportion of this amount is located at United States (78 bbl), Russia (75 bbl), China (32 bbl), Argentina (27 bbl), Libya (26 bbl), United Arab Emirates (23 bbl), Chad (16 bbl), Venezuela (13 bbl) and Mexico (13 bbl) [15]. Typical porosity and permeability of tight oil formations are below 12% and 0.1 mD respectively, though a broader definition of tight oil reservoirs can generally refer to those with very low porosity and permeability [9]. The low- porosity and permeability characteristics furthers the need to stimulate tight oil reservoirs for successful production.

Worldwide, the commercial production of unconventional hydrocarbons is in constant increase. This supplements supply from conventional reservoirs resulting in an overall increase in hydrocarbon production globally and a decrease in prices [16].

This inverse relationship between oil production and oil prices is illustrated in Monge *et al.* [14], where an increase in U. S. oil production from the shale oil boom drives down *West Texas Intermediate* (WTI) oil prices.

Other forms of unconventional gas resource are gas hydrates and coalbed methane reservoirs. Gas hydrates are crystalline ice-like forms of water with a structured molecular framework joined together to create cavities such that gas molecules, which are mostly methane, are trapped within it [17]. Other entrapped guest gases include ethane, isobutene and propane [18]. Natural gas hydrates were only discovered a few decades ago and 98% of deposits occur in upper sedimentary layers underneath the seafloor [7]. It is extensively spread in oceans and polar areas with a reserve that is 10 times greater than global conventional gas [18]. The creation and stability of gas hydrates rely on the properties of both the water and composition of gas, temperature and pressure [18, 19]. The formation of gas hydrate is exothermic, which implies the release of heat during this stage. On the other hand, heat is required for dissociation of hydrates [18–20]. The dissociation of hydrates is an endothermic process relying on the surrounding heat. Gas hydrates are stable at high pressure and low temperature conditions; therefore, depressurisation is an effective means of inducing the release of gas from hydrate deposits [20–22].

Coal seams are dark-banded deposits of coal trapped between layers of rock. They differs from conventional gas reservoirs in terms of their pore structure, porosity, permeability, fluid flow mechanism, gas-water relative permeability and other reservoir characteristics [23]. Coal is both heterogeneous and anisotropic; it is characterised by a dual porosity comprising a porous matrix with micro pores enclosed by a larger scale medium of cleats, which constitute the macro pores [23–25]. Coal porosity and permeability is mostly defined by the micro pores and macro pores, respectively [23]. Usually, water permeates coal seams, which helps to retain the adsorbed gas on the coal surface [25]. Coal seams are unconventional reservoirs containing a variety of gases including methane, hydrogen, ethane, nitrogen and carbon dioxide [26]. It contains a significant proportion of methane, which is more easily extracted in comparison to some of the other gas constituents (e.g., hydrogen and nitrogen). This is due to the reduced affinity coal has for methane. The concentration of methane in the gas content can be as high as 99.95% [27]. The chemical composition of coalbed methane—also known as coal seam gas—is the same as natural gas obtained from conventional reservoirs. The gas is contained in three ways: adsorbed on the surface of micro pores; in a free state in macro pores, i.e., the natural fractures (cleats) within the coal material; and dissolved in the formation water [23, 25, 28].

2. Stimulating strategies and techniques.

2.1 Heavy oil reservoirs

As aptly defined by its name, heavy oil formations contain heavy oils typically characterised by high viscosity and density, and capillarity pressure effects [1, 5]. These peculiar properties make it virtually impossible to exploit heavy oil formations without stimulation. Exploitation can be accomplished by cold production and thermal stimulation. Cold production is a primary recovery method performed at the native reservoir temperature and can achieve a recovery factor between 1 and 10% [5, 29]. This may be carried out by injecting a diluent into the reservoir to reduce the viscosity of the hydrocarbon or by encouraging the initiation and continuous sand production throughout the completion process; the latter is known as *cold heavy oil production with sand* (CHOPS) [5, 30]. Sanding produces high permeability channels referred to as 'wormholes' which enhances recovery [30]. For both approaches, artificial lifts are vital because they lower the producing bottomhole pressure (BHP) thereby increasing the flow rate. Artificial lift systems may consists of pumps (e.g., progressing cavity pumps (PCP) and electrical submersible pumps (ESP)) or a gas lift, whereby injected gas is used to reduce the fluid density of the tubing which is then lifted as the gas expands [31].

Thermal stimulation is an alternative method applied where cold production is not effective or economical. The dependency of oil viscosity on temperature is inverse. Which means it is possible to enhance fluid flow by raising the reservoir temperature. There are several ways this can be realised—for instance, cyclic steam injection and steam flooding [5]. Cyclic steam injection involves two main phases: injection of steam followed by the production of heavy oil with the condensed steam. This is carried out alternatingly with a new cycle started when the rate of oil production declines below a critical level [32]. This method is favoured in the following conditions: in reservoirs that can withstand high-pressure steam and in the presence of thick pay zones (> 10 m) containing sands with high porosity (> 30%) [32]. Steam flooding is the injection of steam into the reservoir to raise the temperature of the oil whilst reducing its viscosity [33, 34]. This method aids the distillation of the light constituents of the oil [35], which further decreases the parent oil viscosity. Steam flooding also reduces the interfacial tension between oil and rock surfaces at the vicinity of the wellbore due to the liberation of the immiscible fluid phase (oil) attached to the host solids by the wetting phase (water) [36].

2.2 Oil shales

Oil shales have very low permeability and porosity. Recovery from such reservoirs can be achieved by the use of horizontal wells [37], hydraulic fracturing, surfactant treatment, matrix acidisation, water imbibition, thermal treatment or a hybrid of these techniques.

2.2.1 Horizontal wells

Horizontal wells have several advantages over vertical wells and are generally more effective in enhancing reservoir performance (**Figure 1**). These include the following: greater and more efficient reservoir drainage and detainment of water production; reduction in gas and water coning; greater rate of production because of the increased reach of the wellbore in the pay zone, since penetration to discrete compartments is possible in complex reservoirs; and reduction in sand production [40, 41]. However, there is a higher cost associated with horizontal wells, which can be up to 2.5 greater than vertical wells [40]. Hence, it is likely that a cost-benefit assessment will be necessary, especially where the increase in reservoir performance is not expected to be intense [41].

2.2.2 Hydraulic fracturing

Hydraulic fracturing is one of the foremost and traditional ways of enhancing fluid flow in oil shales (**Figure 1**). The process induces the initiation and propagation of cracks through the injection of high-pressure fluids with magnitudes that exceed the rock failure stress [42]. The shape, orientation, size and conductivity of the fractures are functions of the direction and magnitude of the formation principal stresses and rock anisotropy, amongst other factors [42]. The in situ principal stress conditions determine the minimum pressure necessary for crack initiation and propagation. The fracturing fluids influence the pattern and behaviour of created fractures. An increasing number of fracturing fluids with

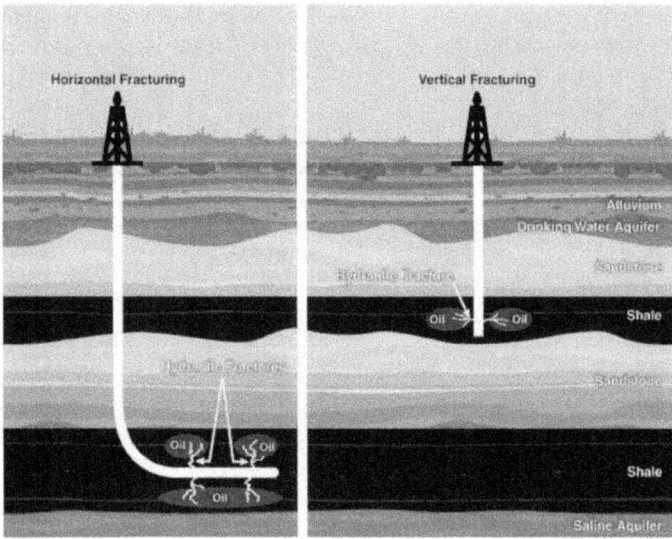

(a)

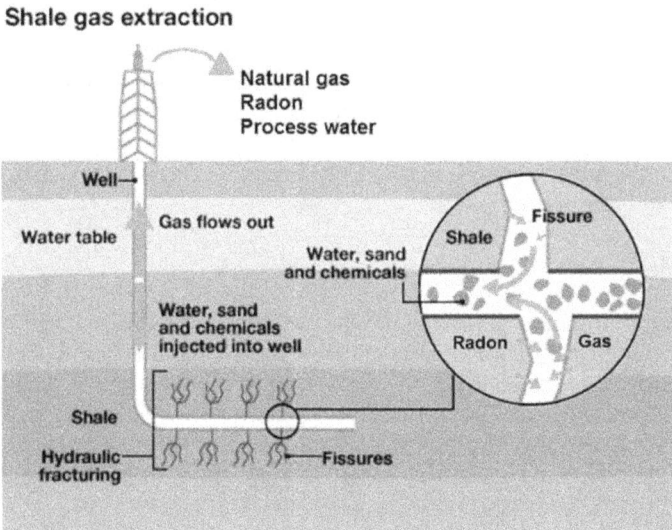

(b)

Figure 1.
Hydraulic fracturing of unconventional reservoirs [38] (a) fracture layout for vertical and horizontal wells [39] (b) fissures created by hydraulic fracturing.

different properties are being demonstrated to be appropriate [42–45] with each category of fluid causing dissimilar effects. This is mainly caused by the differing properties of the fluids. Examples of fracturing fluids include water, CO_2 and oil; they can be generally classified as water-based, oil-based, acid-based and foam-based fluids [46–48]. Fluid density and viscosity are, amongst other primary physical properties, considered when selecting a fracturing fluid. Low-viscosity fracturing fluids produce fractures that are expansive with the tendency to split into several branches. CO_2, for instance, is a low-viscosity fluid that is suitable as a fracturing fluid in oil shales because it creates fractures with surface areas that

are more extensive in comparison to those created by fluids with higher viscosities—e.g., water [44, 49]. High-viscosity fluids such as viscous oils or liquid
CO_2 tend to generate shorter and thicker planar fractures with a small number of
branches [44].

Hydraulic fracturing is not usually applied as a stand-alone strategy to
improve flow in oil shales. It is often used in tandem with other techniques such
as horizontal/inclined wells, thermal dissolution, and the use of special blends
of fracturing fluids like surfactants and other chemicals that aid the recovery of
shale oils [45, 46].

2.2.3 Transverse vertical fractures along horizontal wells

For horizontal wells, the primary recovery method for reservoir stimulation is hydraulic fracturing, whereby transverse fractures that intersect the well
are created (**Figures 2** and **3**). This instigates a substantial pressure drop that
intensifies fluid flow towards the wellbore, thus increasing its performance [52].
Hydraulically fractured horizontal wells perform better than their vertical counterparts (**Figure 2c**). Fractures are orientated either longitudinally or transversely
to the well (**Figure 2a** and **b**). Longitudinal fractures are aligned in the same
direction as the horizontal well; i.e., along the lateral direction parallel to the well
(**Figure 2b**). Horizontal wells with longitudinal fractures are better suited for reservoirs with permeability values that are relatively higher and have a comparable
performance as fractured vertical wells [52–54]. On the contrary, under the same
conditions, transversely fractured horizontal wells perform better in comparison
to both fractured vertical wells and longitudinally fractured horizontal wells
[52–54]. To maximise productivity, the optimal number of transverse fractures
intersecting the horizontal well should be determined; this usually depends on the
fluid and reservoir properties [53].

2.2.4 Surfactant treatment

Surfactants are amphiphilic organic compounds and divided into hydrophobic
and hydrophilic groups. For enhanced oil recovery (EOR), they normally serve as
viscofiers or are used to reduce strong capillary forces in the pores of the reservoir
rock [55, 56]. Oil shales are characterised by their ultra-low permeability. Strong
capillary forces exist in their pores, which hold the oil to the rock surface. To recover
the oil, it is necessary to lessen these capillary forces by altering the interfacial tension, contact angle and wettability [55–57]. Surfactants are used to:

1. Increase the contact angle between the oil liquid-vapour interface and the rock
   surface.

2. Reduce the interfacial or surface tension between different liquids or phases of
   materials (i.e., liquid-liquid, liquid-gas and liquid-solid) (**Figures 4** and **5**).

3. Reduce the oil-wet wettability or in a multiphase (oil-water) fluid system,
   changing it from oil-wet towards water-wet conditions [56, 57] (**Figures 4–6**).

Surfactants are also commonly classified as ionic and non-ionic. Ionic sur-
factants are further categorised as anionic (e.g., Alkyl Aryl Sulfonates, Sodium
Dodecyl Sulfate (SDS) and Alpha-Olefin Sulfonate (AOS)) and cationic (e.g.,
Cetyl Trimethyl Ammonium Bromide (CTAB), Ethoxylated Alkyl Amine and
Dodecyl Trimethyl Ammonium bromide (DTAB)) [61]. Non-ionic surfactants are

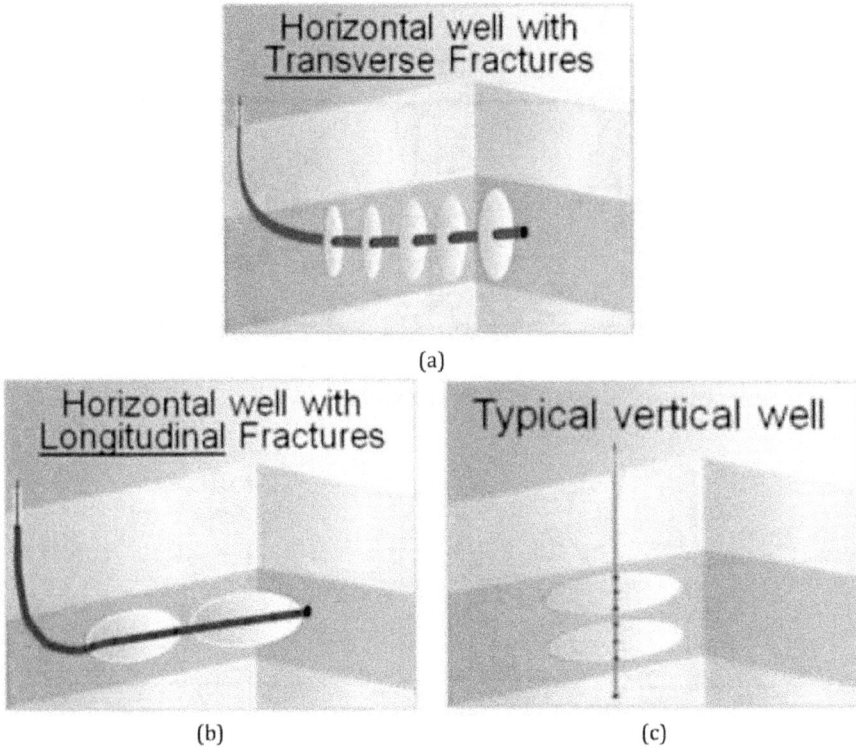

(a)

(b)   (c)

Figure 2.
Configuration of (a) transverse fractures in horizontal well, (b) longitudinal fractures in horizontal well, and (c) fractures in conventional vertical well [50].

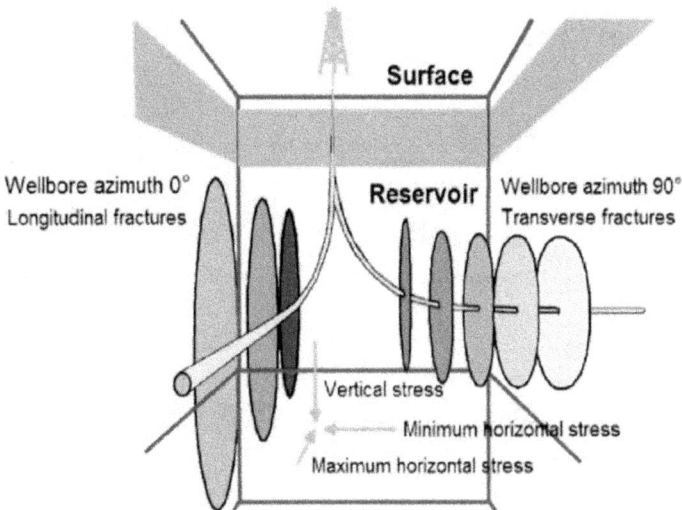

Figure 3.
Relating fracture to horizontal wellbore orientation [51].

not charged; examples of these are Alkyl Polyglycoside (APG), Nonylphenol "N" Ethoxylate and Polyethoxylated Alkyl Phenols) [61]. Other groups of surfactant reported in Negin *et al.* [61] are bio and Zwitterionic surfactants.

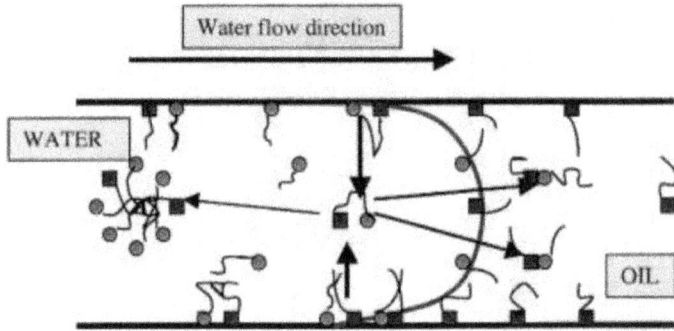

Figure 4.
Mechanism for the alteration of wettability in a pore, from oil-wet to water-wet. Squares are anionic active organic compounds and circles are cationic surfactants [58].

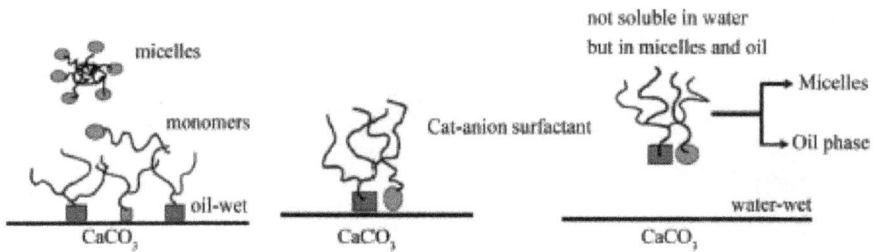

Figure 5.
Mechanism for the alteration of wettability in a rock surface, from oil-wet to water-wet. Circles are cationic surfactants (R-N$^+$ (CH$_3$)$_3$), large squares are crude oil carboxylates and small squares are additional polar compounds [59].

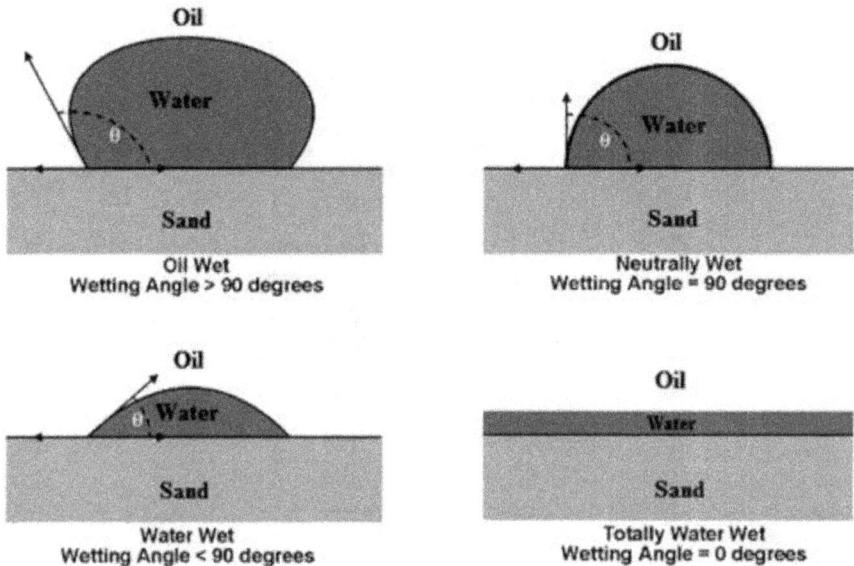

Figure 6.
Alterations in wettability as contact (wetting) angle reduces [60].

2.2.4.1 Wettability

Wettability is the tendency of a fluid to remain in contact with the surface of a solid. For a given wetting fluid, there is an inverse relationship between wettability and contact angle. This means that its wettability decreases when there is a rise in contact angle [62, 63]. The injection of fracturing liquid in the reservoir alters the dynamics of wettability because it introduces another liquid phase to the system. Where two liquids co-exist, one will be wetting and the other non-wetting.

In a multiphase reservoir, such as oil shale, consisting of more than one type of immiscible fluids (e.g., water and shale oil), the wetting fluid preferentially wets the rock surface due its low mobility and stronger attractive forces with the rock. For an oil-water reservoir fluid, water is the denser of the two phases and preferentially wets the rock when the contact angle is less than 90°, the adhesion tension is negative, and the interfacial tension between the water-rock interface exceeds that for the oil-rock interface [64]. The adhesion tension is the difference between the oil-rock and water-rock interfacial tensions. Conversely, oil will be the preferential wetting fluid if the contact angle of water is between 90° and 180°, the adhesion tension is positive, and the interfacial tension between the oil-rock interface exceeds that for the water-rock interface [64]. Water imbibition is boosted as the water-wet wettability increases, resulting in a reduction in the saturation of residual oil [65].

2.2.4.2 Effect of contact angle on wettability

It may not always be easy to define the wettability of a reservoir in a straightforward manner since it is influenced by other factors such as contaminants, surface roughness and time [62, 66]. Nonetheless, the contact angle can serve as a criterion to distinguish between wetting and non-wetting liquids. Whereas, the contact angle of the wetting liquid with the rock is below 90°, for a non-wetting liquid it is between 90° and 180°. If the reservoir consists of both oil and water, the wetting fluid will form a contact angle that is less than 90° [63]. The wetting fluid attaches and spreads along the rock surface thereby enhancing the mobility of the non-wetting fluid. The choice of an appropriate hydraulic fluid should account for this. For instance, water-based fracturing fluids applied in a reservoir will serve as wetting fluids whilst boosting the flow of preexisting hydrocarbons, and the degree of its wettability—in other words, the ease of spread on the rock surface—increases as the contact angle decreases.

2.2.5 Water imbibition

The periodic injection of water into unconventional reservoirs enhances oil recovery because of the imbibition of water by the rock matrix and the displacement of oil trapped within the pores [45]. This technique is fit for shales with a higher water than oil uptake. Shale has a higher affinity for water, which is reflected by larger rates of imbibition [67]. However, it is possible for water blockage to occur resulting in negative impacts on the recovery process [45, 68]. To circumvent this, well-shut operations can be used to drive water further into deeper water-wet sections [45]. Alternatively, surfactants are introduced to improve the water-wet wettability or to completely change the wetting fluid from oil to water [45, 69].

Imbibition is a form of diffusion where a liquid is absorbed into a solid particle resulting in an increase in volume of the particle. It is normally instigated in response to a concentration gradient between the solid (absorbent) and the liquid leading, potentially, to movement of the liquid towards the solid particle. Imbibition is also described as the displacement of an immiscible fluid by another one within a porous medium. This is a typical phenomenon in hydrocarbon reservoirs involving the displacement of the non-wetting fluid out of the pores of the reservoir rock by the wetting fluid [70–72]. It is another means of primary and secondary oil recovery [72]. Water flooding is a form of secondary oil recovery that involves imbibition, where water is injected to displace residual oil in the reservoir [73]. In a water-wet reservoir rock, water—the wetting phase—displaces oil, which is the non-wetting phase [72]. Imbibition is an important process that aid recovery of oil in fractured reservoirs [72, 74, 75].

Imbibition is a complex phenomenon encompassing the multifarious interactions between gravity, capillary and viscous forces. Whereas, gravity and viscous forces are external agents that could be used to drive imbibition, capillary forces are generated internally within the porous medium. On this basis, there are two categories of imbibition: spontaneous/natural and forced. Spontaneous or natural imbibition is the process whereby a wetting fluid displaces a non-wetting fluid within a reservoir rock due to capillary pressure [70, 72, 76, 77]; for instance, water displacing oil in an oil-saturated reservoir rock. On the other hand, forced imbibition are caused by viscous and gravity forces. These external agents create pressure gradients that enable the displacement of non-wetting by wetting fluids. The manner of flow between the wetting and non-wetting fluid determines the type of spontaneous imbibition. Co-current spontaneous imbibition happens where the directions of flow between the wetting and non-wetting fluid are the same. Contrastingly, counter-current spontaneous imbibition happens when the wetting and non-wetting fluid are flowing in opposing directions [70, 75, 78]. In a water-wet reservoir rock, the prevalence of any type of spontaneous imbibition—hence, oil recovery process—depends on the extent of exposure of the rock to water. Oil recovery is dominated by co-current imbibition when the rock is not wholly in contact with water [78]. This form of imbibition is the predominant process that produces oil and occurs in the region of the rock surface in contact with oil. Co-current imbibition evokes a much higher oil recovery rate in comparison to counter-current imbibition, implying a greater production efficiency; in other words, the rock surface in contact with oil produces more oil in contrast to the surface in contact with water [78]. The linear rate of co-current imbibition is shown by Unsal *et al.* [79] to be up to four times higher than counter-current imbibition.

2.2.6 Thermal treatment

Kerogen, which is a solid, insoluble and rich source of organic compounds in oil shale and other sedimentary rocks, can be converted to shale oil by thermal dissolution, hydrogenation or pyrolysis. These are ex situ processes conducted at the ground surface after mining the oil shale and entails the use of very high heat to extract shale oil. Pyrolysis is the thermal decomposition of the organic matter component in solid fuel in an inert environment, and hydrogenation is a chemical treatment involving the reaction between molecular hydrogen and another compound/element with or without the presence of a catalyst. The process can be used to saturate or reduce organic compounds. Hydrogenation can be used to attain high oil yields from oil shales by converting its organic matter content to heavy oil, petrol, etc. [80]. Thermal dissolution is a hydrogen-donor solvent refining process [12]. It is a technique of shale oil extraction, whereby a hydrogen donor solvent such as *tetralin* is introduced into the solid fuel at high temperatures resulting in the depolymerisation, dissolution and cracking of the dissolved organic matter [12, 81, 82].

2.2.7 Acidisation

The injection of certain types of acid into oil shales can lead to rock matrix dis-solution—whereby, for instance, sediments and mud solids are dissolved—increas-ing its permeability and porosity [83]. This technique can be applied to release oil and gas trapped in very small quantities within the rock matrix by repairing a previously damage formation (reflected by a restoration of permeability) and/or enhancing the natural permeability through the creation of additional pores [83]. Examples of acids used in practice are hydrogen chloride (HCL), hydrofluoric acid (HF), and organic forms such as methanoic (formic) acid (HCO_2H or HCOOH) and acetic acid (CH_3COOH). To improve performance, acid blends are frequently used. HCL can be combined with HF or sodium hydroxide (NaOH) [2] or organic acids.

For this technique to be successful, the rock must be, at least, partially soluble in acid. Carbonates are readily soluble in acid; thus, this approach is suitable for carbonate rocks—sedimentary rocks mainly composed of carbonate minerals—e.g., limestone and dolostone [84]. Acidisation is also effectively applied to forma-tions composed primarily of silicate minerals (e.g., sandstone, consisting majorly of aluminosilicates and quartz); however, the two reservoirs (carbonate and silicate reservoirs) are responsive to different types of acids. Sandstones are not soluble in HCL, although this acid is highly acidic. They are more reactive to the relatively weaker HF. HCL is more effective in formations with a rich content of carbonate minerals. Since many formations may be a combination of carbonate and silicate minerals, a blend consisting of a mixture of two or more types of acids is common in practice [2, 45, 83].

Two acidisation techniques are notably used for reservoir stimulation: matrix acidisation and acid fracturing [84]. Matrix acidisation entails the injection of acid into the formation at a pressure below the fracturing point (fracturing pressure). Hence, the formation is not fractured; instead, the acid forming new pathways for fluid flow etches the rock. The key mechanisms include mineral dissolution and the mobilisation of fragmented rock particles resulting in the creation of worm-holes [84].

Acid fracturing is analogous to hydraulic fracturing but with the use of acids to react and etch channels within the walls of the fracture. The central difference between matrix acidisation and acid fracturing is the injection rate. In acid fractur-ing, the solution is pumped into the formation at a high rate leading to a build-up in the fracture pressure, and the initiation and proliferation of fractures. The high flow rate implies that there will be a shorter reaction time and the acid solution is not retained long enough to etch long channels on fracture walls.

Acidisation is less suitable for shale than in other rocks; nonetheless, it can still be applied in stimulating shale formations rich in carbonates [2, 45, 85]. Wormholes are not easily created in shales because of its low permeability, therefore matrix acidisation will likely not be effective [45]. Acid fracturing is the preferred and most suited strategy whereby new fractures are created within the formation and then, together with existing fractures, are roughened by the etching process to fur-ther enhance permeability and porosity. For oil shale formations, further improve-ment in reservoir conductivity is observed through the use of acid blends (e.g., sodium hydroxide mixed with hydrochloric acid (NaOH-HCL) and hydrochloric acid mixed with hydrofluoric acid (HCL-HF)). This is demonstrated in Alhesan *et al.* [2]; however, sufficient enhancement in permeability and porosity can still be established by applying a single type of acid, e.g., HCL, on shales which are rich in carbonates (e.g., [45, 85]). Carbonate minerals such as calcite (calcium carbonate, $CaCO_3$), a constituent of carbonate-rich shale, dissolve in HCL.

Generally, the mineralogy of shale varies between formations and impinges upon its mechanical properties [86, 87]. Shale may content a significant amount of any or a combination of clay, calcite or quartz minerals. Although HCL augments the porosity and permeability of calcite-rick shales, it is observed to have contrary effects on shales with low calcite or high clay content; this is caused by formation damage or impairment as a result of clay swelling and related acid-rock reactions [85]. HCL reaction with calcite is typical presented as [87]:

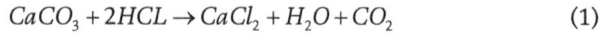

$$CaCO_3 + 2HCL \rightarrow CaCl_2 + H_2O + CO_2 \qquad (1)$$

2.3 Tight reservoirs

Tight oil/gas reservoirs are sometimes referred to as shale reservoirs, but a broader and more accurate definition given in Zhang *et al.* [9] describes it as an ultra-low permeability reservoir rock (sandstone, siltstone, shale and carbonate rocks) closely related to oil shales. The latter concept is adopted in this discourse; notwithstanding, discussions are largely focused on tight sandstones with intermittent allusions to other types of tight oil/gas reservoirs. What qualifies a reservoir to the termed 'tight' is primarily based on its permeability, porosity, and closeness to (or interbedding with) source rocks [9, 88]. Threshold values of 12% for porosity [9] and 0.1 mD for permeability [6, 9, 88] are usually the main distinguishing set of criteria. Recovery from tight reservoirs can be achieved through methods including hydraulic fracturing, water imbibition, surfactant treatment/flooding, acidisation and the generation of an electro-kinetic potential [83, 89–92].

2.3.1 Hydraulic fracturing: tight reservoirs

In a broad sense, the concept of hydraulic fracturing, is generic for all reservoirs, as described in Section 2.2.2. The discussion in this section is not stand-alone; rather, it complements the narrative in Section 2.2.2 and Section 2.2.3. There are three typical approaches for implementing hydraulic fracturing [91, 93]: hydraulic proppant fracturing, water fracturing and hybrid fracturing. The choice of technique is dependent on the formation, and rock and fluid type. Hydraulic proppant fracturing is the conventional technique involving the injection of very viscous gels mixed with a high concentration of proppants. Proppants prop the created fractures thereby maintaining an elevated conductivity. This method creates comparatively short fractures and is suitable for formations of moderate to high permeability [91].

Water fracturing is the injection of water composed of slick water (friction reducers) and a low concentration of proppant to produce extensive but low-width fractures. A conceptual representation of a fracture geometry is illustrated in **Figure 7**. The lengthy geometry of the fracture allows it to connect the wellbore to distant reservoir areas. Water fracturing is appropriate for low permeability (< 1 mD) reservoirs, since fractures with small widths are not effective in moderate to high permeability formations [91, 95, 96]. A key leverage of water fracturing is the considerable cheaper cost in relation to other hydraulic fracturing methods (i.e., hydraulic proppant fracturing and hybrid fracturing), whereas a major weakness is proppant settlement due to the low viscosity of injected fluids, which causes a non-uniform proppant distribution within the propped fracture [95].

Hybrid fracturing is a combination of different hydraulic fracturing stimulation methods, borrowing the advantages of individual treatment approaches. In essence and in the context of the discussion here, it is a blend of hydraulic proppant fracturing and water fracturing. Succinctly, the procedure entails an initial injection

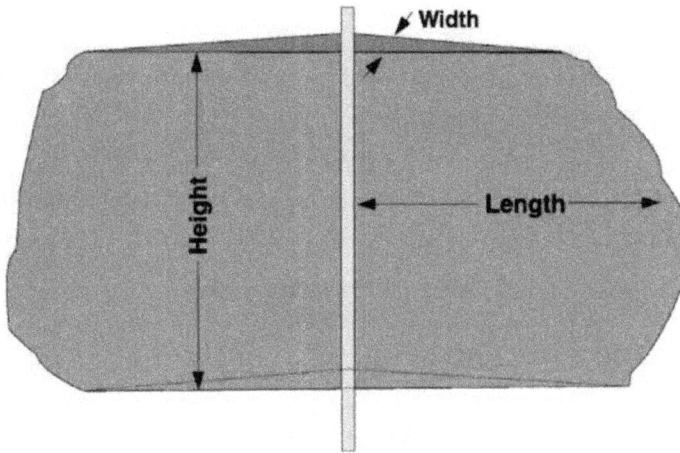

Figure 7.
Fracture geometry as produced by a vertically oriented wellbore [94].

of slick water to create fractures, followed by a treatment with a cross-linked gel consisting of the desired concentration of proppants. The cross-linked gel is conveyed to the extreme ends of the fracture [91]. Hybrid fracturing combines the benefits of both conventional fracturing and water fracturing. Effective fracture half-lengths and fracture conductivities are higher in the induced fractures [93] and the polymer loading in the cross-linked gel is considerably less than what is used for conventional hydraulic proppant fracturing. This has a knock-on effect on the extent of polymer damage [91]. Some of the issues associated with hydraulic prop-pant fracturing are applicable to hybrid treatment [91].

The choice of hydraulic fracturing technique for tight reservoirs depends on several factors. If cost is a chief factor, water fracturing is preferred.

2.3.2 Water imbibition: tight reservoirs

As in oil shales, water imbibition can be employed to enhanced oil and gas recovery in tight reservoirs [90]. Oil is preferentially driven out from pores during water imbibition due to greater capillary forces. Surfactants may be introduced during water imbibition to convert the wettability of rock from *oil-wet* to *water-wet* and to reduce the interfacial tension between liquid phases—oil and water—in the reservoir.

2.3.3 Application of electrokinetics potential

Electrokinetic potential instigates colloidal dispersion. This technique enhances the effect of water imbibition in clay-rich and tight reservoirs (e.g., sandstone) by stimu-lating colloidal movement through the dislodgement and transport of pore clay linings [89]. The removal of clay linings in pores enlarges pore throats and/or creates new flow pathways, causing a direct increase in permeability and porosity. Electrokinetic potential and water injection can be used in tandem to improve efficiency.

2.3.4 Acidisation—Tight reservoirs

Acidisation of carbonate rocks (e.g., limestone and dolostone) to improve per-meability and porosity can be successfully achieved with hydrochloric acid (HCL).

Shales or sandstones containing significant proportions of carbonates can also be treated with HCL. The use of HCL becomes problematic when applied to other kinds of reservoir rocks; for instance, sandstones (chiefly composed of quartz and aluminosilicates), which are insoluble in HCL. The following are some of the problems associated with HCL [83]:

1. It causes formation damage by blocking pore throats thereby reducing porosity and permeability.

2. It escalates the rate of reaction and corrosion at elevated temperatures.

3. There is a risk of later-stage adverse secondary reactions.

However, sandstones react favourably with hydrofluoric acid (HF). The fine particles of quartz and aluminosilicates which block the pores, especially at the near-wellbore region, are soluble in HF. HF can be introduced directly into the reservoir or produced through other chemicals like ammonium bifluoride (NH_4HF_2). Normally, mud acid (a blend of aqueous HCL-HF) is used to repair damages around the vicinity of the wellbore [83, 97]. Fluoride ion (F^-) is the only one of its kind that reacts with quartz in a way that repairs the damage near the wellbore [97]. Mud acid can be prepared by mixing a fluoride ion-releasing chemical, ammonium bi-fluoride salt (NH_4HF_2), with HCL. The reactions are expressed in Eq. (2) [98]:

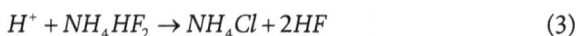

$$HCL \rightarrow H^+ + Cl^- \tag{2}$$

$$H^+ + NH_4HF_2 \rightarrow NH_4Cl + 2HF \tag{3}$$

For sandstone reservoirs, acidisation is performed in three steps: *preflush*, *main flush* and *after flush* [83]. In practice, at the preflush stage, HCL has been used for the dissolution of carbonates and positive ions (e.g., [83, 99–101]); nonetheless, its effectiveness is inconsistent and there are reported incidences of damages [102]. To circumvent this, it is possible to blend HCL with other chemicals to neutralise its adverse effects. This is demonstrated in Shafiq *et al.* [97], where HCL is combined with acetic acid (CH_3COOH) to improve dissolution of carbonates and positive ions (sodium, calcium and magnesium), whilst eluding the damage that would have been triggered by pure HCL.

The second (main) stage of the acidisation process is the use of a fluoride ion (F^-) containing acid to dissolve the minerals (e.g., SiO_2). HF is a commonly used acid but the fluoride ion is very reactive leading to a premature expending of the acid near the wellbore region. To decelerate the reaction rate, HF must be combined with other mineral acids. These are buffer acids, which may be, for instance, HCL or formic acid (HCOOH). The buffer acid retards the reaction rate of HF with the formation and preserves the pH of products of the reaction, which in turn prevents the precipitation of silica [97]. Other acid blends (mud acids) proposed by Shafiq et al., [97] for the main stage of acidisation include Hydrofluoric-phosphoric acid (HF-H₃PO4) and fluoboric-formic acid (H_3OBF_4- HCOOH). In the former, H₃PO4 is a substitute for HCL, while H_3OBF_4 replaces HF in the latter. The product of the reactions between HF and silica mineral is fluosilicic acid (H_2SiF_6). The reaction process is presented in Eqs. (4) and (5) [98, 103]:

$$SiO_2 + 4HF \rightarrow SiF_4 + 2H_2O \tag{4}$$

$$SiF_4 + 2HF \rightarrow H_2SiF_6 \tag{5}$$

Compounds such as silica gelatinous precipitate (Si(OH)$_4$) are eventually formed when fluosilicic acid is decomposed to silicon tetrafluoride (SiF$_4$) (Eq. (6)), which is then hydrolysed (Eq. (7)).

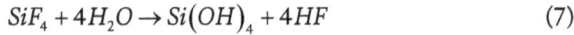

$$H_2SiF_6 \rightarrow 2HF + SiF_4 \tag{6}$$

$$SiF_4 + 4H_2O \rightarrow Si(OH)_4 + 4HF \tag{7}$$

There is also a tendency for other precipitates to form, which can be avoided by the circulation of HCL at the *preflush* stage to remove ions [97, 100]. The *after-flush* stage restores the wettability of the formation and removes the expended acids. Mutual solvents, HCL, acetic acid and other suitable chemicals are candidates for finalising the treatment process [104].

Table 1 is a summary illustration and cross-section of approaches for stimulating the production of unconventional reservoirs.

Reservoir type	Category of stimulation method
Heavy oil formations	Cold production
	• Diluent injection
	• CHOPS
	Thermal stimulation
	• Cyclic steam injection
	• Steam flooding
Oil shale formations	Horizontal wells
	Hydraulic fracturing
	Transverse vertical fractures
	Surfactant treatment
	Water imbibition
	Thermal treatment
	Acidisation
Tight formations	Hydraulic fracturing
	Surfactant treatment
	Water imbibition
	Electro-kinetics potential
	Acidisation

Table 1.
Techniques for stimulating unconventional reservoirs.

3. Fracturing fluids and fluid systems

The crux of hydraulic fracturing is the injection of fluids to generate, within the formation, a pressure that is greater than the breakdown value. The breakdown pressure is fundamentally a function of the formation in situ stresses, the initial pore pressure and the rock tensile strength [105]. Several breakdown pressure models have been developed since the first and classical version derived by Hubbert and Willis [106]. Hubbert and Willis's model is built on the premise that fracture initiation and breakdown takes place when the hoop stress or minimum tangential compressive stress at the wall of the wellbore is equal to the rock tensile strength. The initiated fracture starts to grow when the fracture propagation pressure is attained. For solids-free (clean) injection fluids, the fracture propagation pressure is normally less than that required for fracture initiation [107]. The fracture propagation pressure is the pore pressure at the tip of the fracture. It is lower than the bottom-hole pressure, and this difference depends on permeability, injection rate, fracture length [108], and other factors such as the properties of the fracturing fluid. At times, the fracture propagation pressure is considered as the bottom-hole treating pressure. In this case, its magnitude depends on the in situ stresses and the net drop in pressures [109]. The net pressure drop is influenced by the tortuosity between the wellbore and the fracture, and the viscous flows within the wellbore perforation tunnel and the propagating fracture. The characteristics of the fracturing fluid and fluid system are therefore important in hydraulic fracturing operations.

Some key parameters to consider when choosing or designing a fracturing fluid system include the fluid rheology, conductivity, compatibility between the reservoir rock and fluid, pressure drop along the fracture, environmental impact of the fluid constituents, costs, fluid viscosity and proppant transport ability, and friction losses (in the wellbore, perforations and fractures). The ideal fracturing fluid should be easy to produce; possess enough viscosity for proppant transport and shear resistance; minimise fluid losses, friction forces, and proppant and formation damage; be economically viable; and be compatible with the reservoir rock and in situ fluids [110]. Fracturing fluids can be classified as water-based, oil-based, foam-based and acid-based.

3.1 Water-based fracturing fluids

These are aqueous-based fluids composed of water mixed with proppants and chemical additives such as friction reducers. Water-based fracturing fluids can be categorised as slickwater, linear, crosslinked and viscoelastic surfactant fluids. Slickwater is mainly water; the proportion of water is normally dominant and might be up to 99% of the composition of the fluid. Other constituents (proppants and additives) account for less than 2% of the total volume [110]. The friction reducers (e.g., acrylamide-based polymers, surfactants and biocides) lower the viscosity to values below that for normal water. Because of its low viscosity and proppant concentration, it is possible to inject slickwater at high velocities to create narrow fractures [111].

Linear fluids are uncrosslinked solutions based on polymers (i.e., biopolymers or synthetic polymers or polysaccharides) [111]. Guar, cellulose and their derivatives are examples of biopolymers. Generally, linear fluids are higher in viscosity and thus better than slickwater in proppant conveyance and suspension. Crosslinked solutions are formed when two polymer chains are bonded to enable a fluid type with improved physical characteristics. Examples are crosslinked polymer (guar, guar derivatives, cellulose and cellulose derivatives, etc.) fluids. Typical crosslinkers include borate and other metal-based (Aluminuim, Zirconate, Titanate, etc.) ions [46, 111]. In comparison to linear fluids, crosslinked fluids have higher gel viscosity (hence, proppant carrying capacity) and stability at high temperatures [46, 111]. The high viscosity of crosslinked fluids and their tendency to form filter cakes at the fracture walls means that they must be degraded and removed at the end of the operation using breakers (enzymes and oxidisers), to avoid damage to formation conductivity [46].

Viscous fluids are suitable where high fracture conductivity is desired. Viscoelastic surfactant (VES) fluids are not formed with crosslinkers but mainly reflect the distinctive characteristics of surfactants. They contain less residues and are viscous under shear—i.e., they become highly viscous at low shear rates [46, 112, 113]. For VES fluids, crosslinkers are not necessary; rather, when these fluids are mixed with water the surfactants create micelles that increase its viscosity. Viscosity is reduced when VES fluids are in contact with hydrophobic and organic fluids (e.g., oil and gas); hence, breakers are not required to lower the fluid viscosity during clean-up [46]. VES fluids also facilitate reduction in surface tension of the reservoir fluid, which enables the outflow of water trapped within the rock pores. This is crucial in formations sensitive to water [114]. The deficiencies of VES fluids are their high costs, excessive leak-off rates in very permeable formations (> 200 mD) due to their inability to build a filter cake at the fracture wall, and their instability and decrease in viscosity at high temperatures (> 135°C), [46, 112, 114].

3.2 Oil-based fracturing fluids

Oil-based fracturing fluids are principally applied in formations that are water sensitive. The earliest practices of hydraulic fracturing were conducted using oil-based fracturing fluids [115]. These were mainly hydrocarbons including kerosene, diesel and crude oils. These fluids are generally low in viscosity, which is normally increased by the addition of chemicals such as fatty acids, aluminium phosphates and aluminium esters [114, 116]. An increase in viscosity is imperative for improved stability and proppant-carrying capacity at high temperatures. Oil-based fracturing fluids can perform better than their water-based counterparts. Comparative studies completed by Perfetto *et al.* [116] show that for wells stimulated by oil-based fracturing fluids, there is a slower long-term decline in production, shorter clean-up times and improvements in economic returns. Other notable benefits are minimal contamination, lower specific gravity, lower pipe frictional losses, stability at high shear rates, and less difficulty in optimising proppant pumping and the fracturing process. The drawbacks of oil-based fracturing fluids is the hazard they pose due to high flammability and negative environmental impacts. Additional efforts to combat health and safety concerns are vital.

3.3 Foam-based fracturing fluids

Foam fluids are fundamentally gas/liquid composed of significantly higher proportion of gas in comparison to the liquid fraction. They are viscous fluids. The gas fraction forms the internal phase since it is suspended in the external phase (the liquid). It is differentiated from other gas/liquid mixtures (e.g., dispersions and mist) by the percentage of gas in the total volume. Typically, the gas fraction (F^g) of foams is in the range 52% < = F^g < = 96% [117]. Dispersions (normally classified as energised fluids) and mists consist of gas fractions below 52% and above 96% respectively [114, 117, 118]. **Figures 8** and **9** are schematic representations of these. Foams are characterised by three main parameters: rheology, quality and texture [117]. *Quality*, herein, refers to the percentage of gas in the mixture. The *texture* is the bubble size distribution of the dispersed gas.

Obviously, foams are also preferred for water-sensitive formations because they aid flowback and the amount of water needed for treatment is lower. Carbon dioxide (CO_2) and nitrogen (N_2)—as gas phases—and water, polymers (e.g., guar)

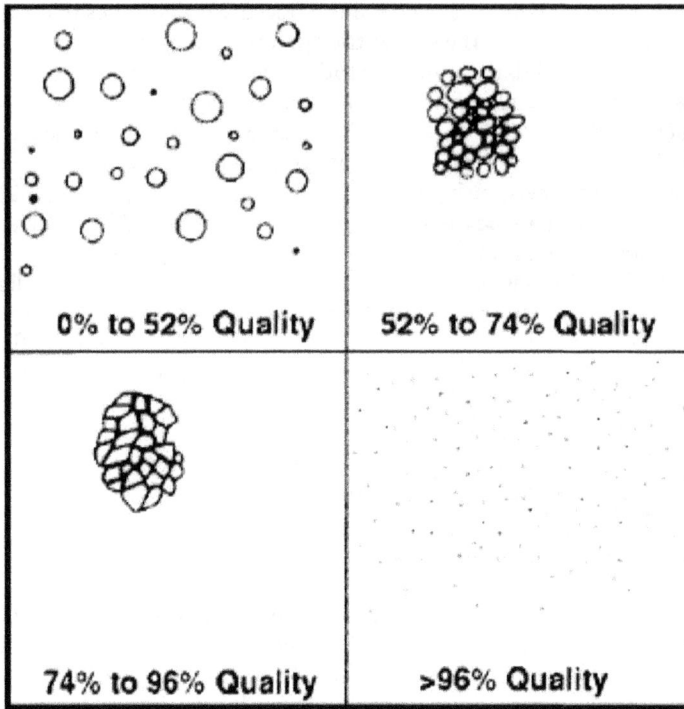

Figure 8.
Foam qualities depicted by different gas–liquid compositions [117].

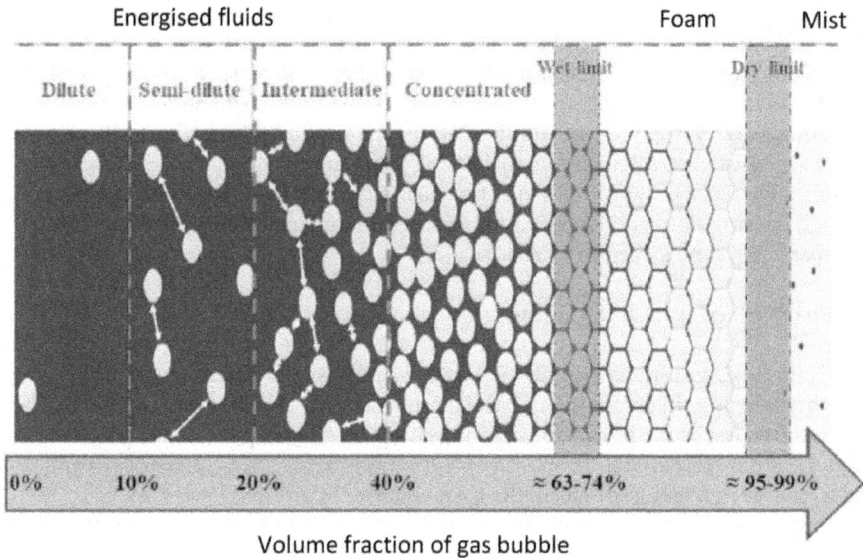

Figure 9.
Classification of gas–liquid mixture depending on the proportion of gas fraction [119].

and acids—as liquid phases—are common components of foams fracturing fluids [120, 121]. **Table 2** presents the main categories of foam-based fracturing fluids. These are water-based, hydrocarbon/oil-based, alcohol-based, acid-based and

Fracturing foam type	Main composition	Target reservoirs
Water-based fracturing foams	Water, foaming surfactant/agent, and N_2 or CO_2 gas	Low pressure formations
Hydrocarbon-based fracturing foams	Hydrocarbon, foaming surfactant/agent and N_2 gas	Water-sensitive low pressure formations
Alcohol-based fracturing foams	Methanol, foaming surfactant/agent, and N_2 gas	Water blocked low pressure formations
Acid-based fracturing foams	Acid, foaming surfactant/agent, CO_2 and N_2 gas	Low pressure and depleted formations
CO2-based fracturing foams	Liquid CO_2, foaming surfactant/agent and N_2 gas	Low pressure formations

Table 2.
Classification and constituents of foam-based fracturing fluids [47, 119, 120, 122, 123].

CO_2-based fracturing foams. Water-based foams are more popular because they are readily available and the technology requirement is low.

The major advantages of foam-based fracturing fluids are as follows [118, 120]:

- Considerable reduction in water requirement in comparison to water-based fracturing fluids; this is directly reflected in the quantity of waste water and its undesirable impact on flora and fauna, and other aspects of the environment

- High recyclability of the foam, which reduces the amount of waste water and cost

- High proppant transporting capacity, which is about 85% greater than water-based fracturing fluids

- Low fluid loss

- Low hydrostatic pressure (head)

- Enablement of backflow of the injected fluid

- Low pressure drops

- Low injection pressure requirements

- Low energy demand for pumping

- Low damage to the formation

- High compatibility with formation fluids

The main disadvantages of foam-based fracturing fluids are given thus [120]:

- Limited choice of surfactants to aid foaming and stabilisation because of the need for them to be completely degradable and environmentally friendly for minimal impact on groundwater, the surrounding land and aquatic life

- High cost due to the peculiarity of equipment, and technical and planning requirements

3.4 Acid-based fracturing fluids

The common types of acid-based fracturing fluids are described in Section 2.2.7 and 2.3.4. These are hydrochloric acid (HCL), hydrofluoric acid (HF), and organic forms of acids such as methanoic (formic) acid (HCO2H or HCOOH) and acetic acid (CH3COOH). During acid fracturing, fracture conductivity is increased by etching channels along fracture walls. This method of fracturing is, therefore, effective in reservoir rocks that are soluble in acid. Carbonate formation rocks (sedimentary) such as limestone and dolostone are soluble in acid; hence, they are the most common beneficiaries of acid fracturing. Nevertheless, innovative applications of acid fracturing on rocks with low acid solubility (e.g., sandstone) are becoming more recognised [83, 97, 124]. HCL is the most popularly used acid fracturing fluid, especially for carbonate rocks, but the solubility of some reservoir rocks (e.g., sandstone) to this acid is low. Thus, the application of HCL in low-soluble formation rock is limited. Although weaker in strength to HCL, HF are more reactive to formation rocks rich in aluminosilicates and quartz—such as sandstone—and, hence, better candidates. In practice, acid blends (mud acids) are preferred and frequently used (e.g., [2, 83, 97, 104]). Examples are HCL-HF, NaOH-HCL, fluoboric-formic acid (H3OBF4-HCOOH) and hydrofluoric-phosphoric acid (HF-H_3PO_4).

4. Summary and conclusion

The imperative of reservoir stimulation is borne out of the need to maximise exploitation of hydrocarbon reserves. Candidate reservoir formations that benefit from stimulation operations span across both conventional and unconventional reservoirs. Stimulation is necessary in conventional reservoirs to enhance the productivity of depleted oil and gas formations, which is accomplished through enhanced oil/gas recovery (EOR & EGR) strategies. EOR/EGR is even more pertinent to the exploitation of unconventional reservoirs (i.e., heavy oil, oil shales, tight sandstones, tight limestone formations, etc.). The advent of the exploration of these peculiar hydrocarbon formations has revolutionised the oil and gas industry, driving down energy prices and revealing potential opportunities for cleaner fuels. It is also essential for coal bed methane (CBM) reservoirs to be stimulated in order to instigate and/or improve productivity.

Unconventional reservoirs are complex, distinctive and diverse. They greatly differ from conventional reservoirs in terms of their structure, composition, and rock and fluid properties. Due to these features, there are increased challenges in producing from this type of reservoirs. The stimulation of oil and gas unconventional reservoirs has been ongoing for many decades and over the years, the employed techniques have evolved to become more effective, economical, sustainable and environment-friendly. The diversity of unconventional reservoirs with respect to their structural layout, rock type, hydrocarbon content, proximity to conventional formations, etc., brings to the fore the impracticability of applying a single set of stimulating techniques across board.

Heavy oil reservoirs consist of high-viscosity and high-density hydrocarbon fluids. They are generally produced via two methods: cold production and thermal stimulation. Cold production is carried out either by injecting a diluent to decrease the viscosity of the reservoir fluid or by 'cold heavy oil production with sand' (CHOPS). Alternatively, thermal stimulation is typically implemented in any of the following two ways: cyclic steam injection and steam flooding. Oil shale reservoirs are normally produced by any or a combination of methods, including the use of horizontal wells, hydraulic fracturing, creating transverse vertical fractures (along horizontal wells), surfactant treatment/flooding, water imbibition, thermal treatment and acidisation. Tight reservoirs are primarily produced by hydraulic fracturing, surfactant treatment/flooding, water imbibition, application of electrokinetic potential and acidisation.

The process of selecting an appropriate approach is an important aspect of the routine because of the disparity in different types of unconventional reservoirs and the availability of a seemingly wide range of options of stimulating techniques. A thorough site reconnaissance and an assessment of the effectiveness, efficiency and impact of the selected method is indispensable. These should consider, amongst other factors, reservoir productivity, cost, environmental impact, and health and safety.

Author details

Kenneth Imo-Imo Israel Eshiet
Faculty of Science and Engineering, School of Architecture and Built Environment, University of Wolverhampton, UK

*Address all correspondence to: kenieshiet@yahoo.com

IntechOpen

References

[1] Alajmi HM. Assessment of Development Methods for a Heavy Oil Sandstone Reservoir. London: Imperial College; 2013

[2] Alhesan JSA, Amer MW, Marshall M, Jackson WR, Gengenbach T, Qi Y, et al. A comparison of the NaOH-HCl and HCl-HF methods of extracting kerogen from two different marine oil shales. Fuel. 2019;**236**:880-889

[3] Briggs PJ, Baron RP, Fulleylove RJ, Wright MS. Development of heavy-oil reservoirs. Journal of Petroleum Technology. 1988;**40**(2):206-213

[4] Chopra S, Lines LR, Schmitt DR, Batzle M. Heavy-oil reservoirs: Their characterization and production. In: Chopra S, Lines LR, Schmitt DR, Batzle ML, editors. Heavy Oils: Reservoir Characterization and Production Monitoring. Geophysical Developments No. 13. Houston: Society of Exploration Geophysicists. Houston: Society of Exploration Geophysicists. 2010. p. 1-69. DOI: 10.1190/1.9781560802235.ch1. ISBN (online): 978-1-56080-223-5

[5] Curtis C, Kopper R, Decoster E, Guzman-Garcia A, Huggins C, Knauer L, et al. Heavy-oil reservoirs. Oilfield Review. 2002;**14**(3)

[6] Holditch SA. Tight gas sands. Paper SPE-103356 of distinguish author series. Journal of Petroleum Technology. 2006;**58**(6):86-93

[7] Makogon YF, Perry KF, Holste JC. Gas hydrate deposits: Formation and development. In: Paper presented at the Offshore Technology Conference. Houston, Texas: Offshore Technology Conference; 2004

[8] Dyni JR. Geology and resources of some world oil-shale deposits. U.S. Geological Survey Scientific Investigations Report 2005-5294. 2006. Available from: https://pubs.usgs.gov/sir/2005/5294/pdf/sir5294_508.pdf [Accessed: 01 February 2022]

[9] Zhang X-S, Wang H-J, Ma F, Sun X-C, Zhang Y, Song Z-H. Classification and characteristics of tight oil plays. Petroleum Science. 2016;**13**(1):18-33

[10] Youngquist W. Shale oil - The elusive energy. Hubbert Center Newsletter 98/4. M. King Hubbert Center for Petroleum Supply Studies, Colorado Sch. Mines, Golden Colorado. Colorado: Center for Petroleum Supply Studies. 1998. p. 1-8. Available online: https://energyskeptic.com/wp-content/uploads/2011/07/Youngquist_Shale-oil.pdf [Accessed 16 August 2022]

[11] Biglarbigi K, Crawford P, Carolus M, Dean C. Rethinking world oil-shale resource estimates. In: Paper presented at the SPE Annual Technical Conference and Exhibition. Florence, Italy: Society of Petroleum Engineers (SPE); 2010. Paper Number: SPE-135453-MS. DOI: 10.2118/135453-MS

[12] Gorlov EG. Thermal dissolution of solid fossil fuels. Solid Fuel Chemistry. 2007;**41**:290

[13] Luik H. Alternative technologies for oil shale liquefaction and upgrading. In: Paper Presented at the International Oil Shale Symposium; Tallinn, Estonia: Estonian Academy Publishers; 8-11 June 2009

[14] Monge M, Gil-Alana L, Perez de Gracia F. U.S. shale oil production and WTI prices behaviour. Energy. 2017;**141**:12-19

[15] EIA. World shale resource assessments. Analysis and projections. 2015. Available from: https://www.eia.gov/analysis/studies/worldshalegas/ [Accessed: January 2019]

[16] Manescu CB, Nuno G. Quantitative effects of the shale oil revolution. Energy Policy. 2015;**86**:855-866

[17] Rajput S, Thakur NK. Chapter 2—Generation of methane in earth. In: Geological Controls for Gas Hydrates Formations and Unconventionals. Amsterdam: Elsevier; 2016

[18] Zou C. Unconventional Petroleum Geology. Amsterdam: Elsevier; 2013. pp. 337-354

[19] Carlisle TR, Hodgson GW. The formation of natural-gas hydrates in sedimentary rock. Chemical Geology. 1985;**49**:371-383

[20] Wang P, Wang S, Song Y, Yang M. Methane hydrate formation and decomposition properties during gas migration in porous medium. Energy Procedia. 2017;**105**:4668-4673

[21] Li B, Li XS, Li G, Feng JC, Wang Y. Depressurization induced gas production from hydrate deposits with low gas saturation in a pilot-scale hydrate simulator. Applied Energy. 2014;**129**:274-286

[22] Song Y, Cheng C, Zhao J, Zhu Z, Liu W, Yang M, et al. Evaluation of gas production from methane hydrates using depressurization, thermal stimulation and combined methods. Applied Energy. 2015;**145**:265-277

[23] Guo X, Du Z, Li S. Computer modelling and simulation of coalbed methane reservoir. In: Paper presented at the SPE Eastern Regional/AAPG Eastern Section Joint Meeting. Pittsburgh, Pennsylvania: Society of Petroleum Engineers (SPE); 2003. Paper Number: SPE-84815-MS. DOI: 10.2118/84815-MS

[24] Lu M, Connell L. Dual porosity processes in coal seam reservoirs. In: Paper presented at the SPE Asia Pacific Oil and Gas Conference and Exhibition. Brisbane, Queensland, Australia: Society of Petroleum Engineers (SPE); 2010. Paper Number: SPE-133100-MS. DOI: 10.2118/133100-MS

[25] Rajput S, Thakur NK. Chapter 7—Reservoir characteristics. In: Geological Controls for Gas Hydrates Formations and Unconventionals. Amsterdam: Elsevier; 2016

[26] Kim AG. The Composition of Coalbed Gas. Washington: Bureau of Mines; 1973

[27] Li M, Jiang B, Lin S, Lan F, Zhang G. Characteristics of coalbed methane reservoirs in Faer coalfield, Western Guizhou. Energy Exploration and Exploitation. 2013;**31**(3):411-428

[28] Logan TL. Horizontal drainhole drilling techniques used for coal seam resource exploitation. In: Paper presented at the SPE Annual Technical Conference and Exhibition. Houston, Texas: Society of Petroleum Engineers (SPE); 1988. Paper Number: SPE-18254-MS. DOI: 10.2118/18254-MS

[29] Temblay B. Chapter 22—Cold production of heavy oil. In: Sheng JJ, editor. Enhanced Oil Recovery—Field Case Studies. Cambridge, Massachusetts: Gulf Professional Publishing, Elsevier; 2013. pp. 615-666

[30] Haddad AS, Gates I. Modelling of cold heavy oil production with sand (CHOPS) using a fluidized sand algorithm. Fuel. 2015;**158**:937-947

[31] Worth D, Al-Safran E, Choudhuri A, Al-Jasmi A. Assessment of artificial lift methods for a heavy oil field in Kuwait. In: Paper presented at the SPE International Heavy Oil Conference and Exhibition. Mangaf, Kuwait: Society of Petroleum Engineers (SPE); 2014. Paper Number: SPE-172883-MS. DOI: 10.2118/172883-MS

[32] Speight JG. Chapter 5—Thermal methods of recovery. In: Heavy Oil

Production Processes. Cambridge, Massachusetts: Gulf Professional Publishing, Elsevier; 2013. pp. 93-130

[33] Fanchi JR. Enhanced recovery and coal gas modeling. In: Principles of Applied Reservoir Simulation. Fourth ed. Cambridge, Massachusetts: Gulf Professional Publishing, Elsevier; 2018

[34] Martin FD, Colpitts RM. Reservoir Engineering. In: Lyons WC, editor. Standard Handbook of Petroleum and Natural Gas Engineering. Woburn, Massachusetts: Butterworth-Heinemann Elsevier; 1996;**2**

[35] Mollaei A, Maini B. Steam flooding of naturally fractured reservoirs: Basic concepts and recovery mechanisms. Journal of Canadian Petroleum Technology. 2010;**49**:1

[36] He L, Lin F, Li X, Sui H, Xu Z. Interfacial sciences in unconventional petroleum production: From fundamentals to applications. Chemical Society Reviews. 2015;**00**:1-3

[37] Chaudhary AS, Ehlig-Economides C, Wattenbarger R. Shale oil production performance from a stimulated reservoir volume. In: Paper Presented at the SPE Annual Technical Conference and Exhibition; Denver, Colorado, USA; October 2011. Paper Number: SPE-147596-MS. 2011. DOI: 10.2118/147596-MS

[38] Carvalho F. Mining industry and sustainable development: Time for change. Food and Energy Security. 2017;**6**(2):61-77. DOI: 10.1002/fes3.109

[39] LAO. Hydraulic Fracturing: How it Works and Recent State Oversight Actions. California, USA: Legislative Analyst's Office; 2016

[40] Joshi SD. Cost/benefits of horizontal wells. In: Paper presented at the SPE Western Regional/AAPG Pacific Section Joint Meeting. Long Beach,

California: Society of Petroleum Engineers (SPE); 2003. Paper Number: SPE-83621-MS. DOI: 10.2118/83621-MS

[41] Schlumberger. Middle East Well Evaluation Review: Horizontal Highlights. Schlumberger: Middle East & Asia Reservoir Review; 1996

[42] Bennour Z, Ishida T, Nagaya Y, Chen Y, Nara Y, Chen Q, et al. Crack extension in hydraulic fracturing of shale cores using viscous oil, water, and liquid carbon dioxide. Rock Mechanics and Rock Engineering. 2015;**48**:1463-1473. DOI: 10.1007/s00603-015-0774-2

[43] Ishida T, Chen Q, Mizuta Y, Roegiers J-C. Influence of fluid viscosity on the hydraulic fracturing mechanism, transactions of the ASME. Journal of Energy Resources Technology. 2004;**126**:190-200

[44] Ishida T, Nagaya Y, Inui S, Aoyagi K, Nara Y, Chen Y, et al. AE monitoring of hydraulic fracturing experiments with CO_2 and water. In: Proceedings of Eurock2013. Wroclaw, Poland: CRC Press; 2013. pp. 957-962. Paper Number: ISRM-EUROCK-2013-149. ISBN-13: 978-1-138-00080-3

[45] Miller C, Tong S, Mohanty KK. A chemical blend for stimulating production in oil-shale formations. In: Paper presented at the SPE/AAPG/SEG Unconventional Resources Technology Conference. Houston, Texas, USA: Unconventional Resources Technology Conference; 2018. Paper Number: URTEC-2900955-MS. DOI: 10.15530/URTEC-2018-2900955

[46] Barati R, Liang J-T. A review of fracturing fluid systems used for hydraulic fracturing of oil and gas wells. Journal of Applied Polymer Science. 2014;**131**(16):1-11

[47] Gandossi L. An Overview of Hydraulic Fracturing and Other Formation Stimulation Technologies for

Shale Gas Production. Joint Research Centre of the European Commission. Publications Office of the European Union; 2013. JRC86065. ISBN: 978-92-79-34729-0. DOI: 10.2790/99937

[48] Montgomery, C. (2013) Fracturing fluids components. In Effective and Sustainable Hydraulic Fracturing, eds. A.P. Bunger, J. McLennan, and R. Jeffrey. London: INTECH;

[49] Ishida T, Aoyagi K, Niwa T, Chen Y, Murata S, Chen Q, et al. Acoustic emission monitoring of hydraulic fracturing laboratory experiment with supercritical and liquid CO2. Geophysical Research Letters. 2012;**39**:L16309. DOI: 10.1029/2012GL052788

[50] Allan ME, Gold DK, Reese DW. Development of the Belridge field's diatomite reservoirs with hydraulically fractured horizontal wells: From first attempts to current ultra-tight spacing. In: Paper presented at SPE Annual Technical Conference and Exhibition. Florence, Italy: Society of Petroleum Engineers; 2010

[51] Lui P, Feng Y, Zhao L, Li N, Luo Z. Technical status and challenges of shale gas development in Sichuan Basin, China. Petroleum. 2015;**1**(1):1-7

[52] Wei Y, Economides MJ. Transverse hydraulic fractures from a horizontal well. In: Paper presented at the SPE Annual Technical Conference and Exhibition. Dallas, Texas: Society of Petroleum Engineers (SPE); 2005. Paper Number: SPE-94671-MS. DOI: 10.2118/94671-MS

[53] Soliman MY, Hunt JL, Azari M. Fracturing horizontal wells in gas reservoirs. SPE Production and Facilities. 1999;**14**(4):277-283

[54] Villegas ME, Wattenbarger RA, Valkó P, Economides MJ. Performance of longitudinally fractured horizontal wells in high-permeability anisotropic formations. In: Paper presented at SPE

Annual Technical Conference and Exhibition. Denver, Colorado: Society of Petroleum Engineers (SPE); 1996. Paper Number: SPE-36453-MS. DOI: 10.2118/36453-MS

[55] Abdulelah H, Mahmood SM, Al-Mutarreb A. Effect of anionic surfactant on wettability of shale and its implication on gas adsorption/desorption behavior. Energy & Fuels. 2018;**32**(2):1423-1432

[56] Mirchi V, Saraji S, Goual L, Piri M. Dynamic interfacial tensions and contact angles of surfactant-in-brine/oil/shale systems: Implications to enhanced oil recovery in shale oil reservoirs. In: Presentation at the SPE Improved Oil Recovery Symposium. Tulsa, Oklahoma, USA: Society of Petroleum Engineers (SPE); 2014. Paper Number: SPE-169171-MS. DOI: 10.2118/169171-MS

[57] Mirchi V, Saraji S, Goual L, Piri M. Experimental investigation of surfactant flooding in shale oil reservoirs: Dynamic interfacial tension, adsorption and wettability. In: Paper presented at the SPE/AAPG/SEG Unconventional Resources Technology Conference (URTeC). Denver, Colorado, USA: Unconventional Resources Technology Conference; 2014. Paper Number: URTEC-1913287-MS. DOI: 10.15530/URTEC-2014-1913287

[58] Standnes DC, Austad T. Wettability alteration in chalk: 2. Mechanism for wettability alteration from oil-wet to water-wet using surfactants. Journal of Petroleum Science and Engineering. 2000;**28**:123-143. DOI: 10.1016/S0920-4105(00)00084-X

[59] Sheng J. Review of surfactant enhanced oil recovery in carbonate reservoirs. Advances in Petroleum Exploration and Development. 2013;**6**(1):1-10

[60] Kantzas A, Bryan J, Taheri S. Fundamentals of Fluid Flow in Porous Media. Calgary: PERM Inc; 2019

[61] Negin C, Ali S, Xie Q. Most common surfactants employed in chemical enhanced oil recovery. Petroleum. 2017;3(2):197-211. DOI: 10.1016/j.petlm.2016.11.007

[62] Ksiezniak K, Rogala A, Hupka J. Wettability of shale rock as an indicator of fracturing fluid composition. Physicochemical Problems of Mineral Processing. 2015;51(1):315-323

[63] Schön JH. Physical Properties of Rocks: Fundamentals and Principles of Petrophysics. Second Edition. Physical Properties of Rocks: Developments in Petroleum Science. Amsterdam: Elsevier; 2015;65

[64] Crain ER. Chapter 9: Lab data acquisition. In: Crain's Petrophysical Handbook. 2010

[65] Sheng JJ. Status of surfactant EOR technology. Petroleum. 2015;1(2):97-105

[66] Zhang Y, Li Z, Lai F, Wu H, Mao G, Adenutsi CD. Experimental investigation into the effects of fracturing fluid-shale interaction on pore structure and wettability. Geofluids. 2021;2021:6637955

[67] Ghanbari E, Dehghanpour H. Impact of rock fabric on water imbibition and salt diffusion in gas shales. International Journal of Coal Geology. 2015;138:55-67. DOI: 10.1016/j.coal.2014.11.003

[68] Dehghanpour H, Xu M, Habibi A. Wettability of gas shale reservoirs. In: Rezaee R, editor. Fundamentals of Gas Shale Reservoirs. 2015. pp. 341-359

[69] Zeng T, Miller CS, Mohanty K. Application of surfactants in shale chemical EOR at high temperatures. In: Paper presented at the SPE Improved Oil Recovery Conference. Tulsa, Oklahoma: Society of Petroleum Engineers (SPE); 2018. Paper Number: SPE-190318-MS. DOI: 10.2118/190318-MS

[70] Abd AS, Elhafyan E, Siddiqui AR, Alnoush W, Blunt MJ. A review of the phenomenon of counter-current spontaneous imbibition: Analysis and data interpretation. Journal of Petroleum Science and Engineering. 2019;180:456-470

[71] Lu NB, Amchin DB, Datta SS. Forced imbibition in stratified porous media: Fluid dynamics and breakthrough saturation. Physical Review Fluids. 2021;6(11):114007

[72] Patel HS, Meher R. Simulation of counter-current imbibition phenomenon in a double phase flow through fracture porous medium with capillary pressure. Ain Shams Engineering Journal. 2018;9:2163-2169

[73] Vishnyakov V, Suleimanov B, Salmanov A, Zeynalov E. Chapter: 7 oil recovery stages and methods. In: Vishnyakov V, Suleimanov B, Salmanov A, Zeynalov E, editors. Primer on Enhanced Oil Recovery. Cambridge, Massachusetts: Gulf Professional Publishing, Elsevier; 2020

[74] Morrow NR, Mason G. Recovery of oil by spontaneous imbibition. Current Opinion in Colloid & Interface Science. 2001;6(4):321-337

[75] Unsal E, Mason G, Ruth DW, Morrow NR. Co- and counter-current spontaneous imbibition into groups of capillary tubes with lateral connections permitting cross-flow. Journal of Colloid and Interface Science. 2007;315:200-209

[76] Zahasky C, Benson SM. Spatial and temporal quantification of spontaneous imbibition. Geophysical Research Letters. 2019;46:11972-11982

[77] Zhang L, Wang K, An H, Li G, Su Y, Zhang W, et al. Spontaneous imbibition of capillaries under the end effect and wetting hysteresis. ACS Omega. 2022;7(5):4363-4371

[78] Pooladi-Darvish M, Firoozabadi A. Cocurrent and countercurrent

imbibition in a water-wet matrix block. SPE Journal. 2000;**5**(1):3-11

[79] Unsal E, Mason G, Morrow NR, Ruth DW. Co-current and counter-current imbibition in independent tubes of non-axisymmetric geometry. Journal of Colloid and Interface Science. 2007;**306**(1):105-117

[80] Pier M. Treatment of shale oil by hydrogenation. In: Paper presented at the Conference on Oil Shale and Cannel Coal. Glasgow, London: Institute of Petroleum; 1938

[81] Johannes I, Tiikma L, Luik H, Tamvelius H, Krasulina J. Catalytic thermal liquefaction of oil shale in tetralin. ISRN Chemical Engineering. 2012;**2012**:11

[82] Purevsuren B, Ochirbat P. Comparative study of pyrolysis and thermal dissolution of Estonian and Mongolian Khoot oil shales. Oil Shale. 2016;**33**(4):329-339

[83] Shafiq MU, Mahmud HKB, Rezaee R. New acid combination for a successful sandstone acidizing. In: IOP Conference Series: Materials Science and Engineering. 2017

[84] Khadeeja A. Acidizing Oil Wells, a Sister-Technology to Hydraulic Fracturing: Risks, Chemicals, and Regulations. Los Angeles: University of California; 2016

[85] Teklu TW, Park D, Jung H, Amini K, Abass H. Effect of dilute acid on hydraulic fracturing of carbonate-rich shales: Experimental study. SPE Production and Operation. 2017;**34**(01):170-184. DOI: 10.2118/187475-PA

[86] Morsy S, Sheng JJ, Hetherington CJ, Soliman MY, Ezewu RO. Impact of matrix acidizing on shale formations. In: Paper presented at the SPE Nigeria Annual International Conference and Exhibition. August, Lagos, Nigeria: Society of Petroleum Engineers (SPE); 2013. Paper Number: SPE-167568-MS. DOI: 10.2118/167568-MS

[87] Patton BJ, Pitts F, Goeres T, Hertfelder G. Matrix acidizing case studies for the point arguello field. In: Paper presented at the SPE Western Regional/AAPG Pacific Section Joint Meeting. Long Beach, California: Society of Petroleum Engineers (SPE); 2003. Paper Number: SPE-83490-MS. DOI: 10.2118/83490-MS

[88] Jia C, Zou C, Li J, Li D, Zheng M. Evaluation criteria, major types, characteristics and resource prospects of tight oil in China. Petroleum Research. 2016;**1**(1):1-9

[89] Alklih MY, Ghosh B, Al-Shalabi EW. Tight reservoir stimulation for improved water injection—A novel technique. In: Paper presented at International Petroleum Technology Conference. Doha, Qatar: International Petroleum Technology Conference; 2014. Paper Number: IPTC-17536-MS. DOI: 10.2523/IPTC-17536-MS

[90] Cheng Z, Wang Q, Ning Z, Li M, Lyu C, Huang L, et al. Experimental investigation of countercurrent spontaneous imbibition in tight sandstone using nuclear magnetic resonance. Energy and Fuels. 2018;**32**(6):6507-6517. DOI: 10.1021/acs.energyfuels.8b00394

[91] Reinicke A, Rybacki E, Stanchits S, Huenges E, Dresen G. Hydraulic fracturing stimulation techniques and formation damage mechanisms—Implications from laboratory testing of tight sandstone–proppant systems. Chemie der Erde. 2010;**70**(S3):107-117

[92] Jing W, Huiqing L, Genbao Q, Yongcan P, Yang G. Investigations on spontaneous imbibition and the influencing factors in tight oil reservoirs. Fuel. 2019;**236**:755-768. DOI: 10.1016/j.fuel.2018.09.053

[93] Rushing JA, Sullivan RB. Evaluation of hybrid water-Frac stimulation technology in the bossier tight gas sand play. In: Paper presented at the SPE Annual Technical Conference and Exhibition. Denver, Colorado: Society of Petroleum Engineers (SPE); 2003. Paper Number: SPE-84394-MS. DOI: 10.2118/84394-MS

[94] Daneshy A. Hydraulic fracturing to improve production. The way ahead. Society of Petroleum Engineers. 2010;6(3):14-17

[95] Britt LK, Smith MB, Haddad Z, Lawrence P, Chipperfield S, Hellmann T. Water-fracs: We do need proppant after all. In: SPE102227. 2006

[96] Mayerhofer MJ, Richardson MF, Walker RN Jr, Meehan DN, Oehler MW, Browing RR Jr. Proppants? We don't need no proppants. In: Paper presented at the SPE Annual Technical Conference and Exhibition. San Antonio, Texas: Society of Petroleum Engineers (SPE); 1997. Paper Number: SPE-38611-MS. DOI: 10.2118/38611-MS

[97] Shafiq MU, Mahmud HKB, Rezaee R. An effective acid combination for enhanced properties and corrosion control of acidizing sandstone formation. In: IOP Conference Series: Materials Science and Engineering. 2016

[98] Al-Harbi BG, Al-Khaldi MH, Al-Dossary KA. Interactions of organic-HF systems with aluminosilicates: Lab testing and field recommendations. In: Society of Petroleum Engineers: SPE European Formation Damage Conference. Noordwijk, the Netherlands: Society of Petroleum Engineers (SPE); 2011. Paper Number: SPE-144100-MS. DOI: 10.2118/144100-MS

[99] Gdanski RD. Kinetics of the primary reaction of HF on Alumino-silicates. SPE Production and Facilities. 2000;15(4):279-287

[100] Hill AD, Sepehrnoori K. Design of the HCl preflush in sandstone acidizing. SPE Production & Facilities. 1994;9(2): 115-120. DOI: 10.2118/21720-PA

[101] Ji Q, Zhou L, Nasr-El-Din HA. Acidizing sandstone reservoirs using fines control acid. In: Paper presented at the SPE Latin America and Caribbean Petroleum Engineering Conference. Maracaibo, Venezuela: Society of Petroleum Engineers (SPE); 2014. Paper Number: SPE-169395-MS. DOI: 10.2118/169395-MS

[102] Thomas RL, Nasr-El-Din HA, Mehta S, Hilab V, Lynn JD. The impact of HCl to HF ratio on hydrated silica formation during the acidizing of a high temperature sandstone gas reservoir in Saudi Arabia. In: Paper presented at SPE Annual Technical Conference and Exhibition. San Antonio, Texas: Society of Petroleum Engineers (SPE). 2002. Paper Number: SPE-77370-MS. DOI: 10.2118/77370-MS

[103] Shaughnessy CM, Kunze KR. Understanding sandstone acidizing leads to improved field practices. Journal of Petroleum Technology. 1981;33(7):1196-1202

[104] Shafiq MU, Mahmud HB. Sandstone matrix acidizing knowledge and future development. Journal of Petroleum Exploration and Production Technology. 2017;7(4):1205-1216

[105] Guo F, Morgenstern NR, Scott JD. Interpretation of hydraulic fracturing breakdown pressure. International Journal of Rock Mechanics and Minining Sciences. 1993;30(6):617-626

[106] Hubbert KM, Willis DG. Mechanics of hydraulic fracturing. Petroleum Transactions on AIME. 1957;210:153-166

[107] Feng Y, Jones JF, Gray KE. A review on fracture-initiation and -propagation pressures for lost circulation and

wellbore strengthening. SPE Drilling and Completion. 2016;**31**(2):134-144. DOI: 10.2118/181747-PA

[108] Sun J, Deng J, Yu B, Peng C. Model for fracture initiation and propagation pressure calculation in poorly consolidated sandstone during waterflooding. Journal of Natural Gas Science and Engineering. 2015;**22**:279-291

[109] Belyadi H, Fathi E, Belyadi F. Hydraulic fracturing in unconventional reservoirs. In: Theories, Operations, and Economic Analysis. Chapter 9-Fracture Pressure Analysis and Perforation Design. Cambridge, Massachusetts: Gulf Professional Publishing (Elsevier); 2017

[110] King GE. Hydraulic fracturing 101: What every representative, environmentalist, regulator, reporter, investor, university researcher, neighbor and engineer should know about estimating frac risk and improving frac performance in unconventional gas and oil wells. In: Paper presented at SPE Hydraulic Fracturing Technology Conference. The Woodlands, Texas, USA: Society of Petroleum Engineers (SPE); 2012. Paper Number: SPE-152596-MS. DOI: 10.2118/152596-MS

[111] Li L, Al-Muntasheri GA, Liang F. A review of crosslinked fracturing fluids prepared with produced water. Petroleum. 2016;**2**(4):313-323

[112] Al-Ghazal M, Al-Driweesh S, Al-Shammari F. First successful application of an environment friendly fracturing fluid during on-the-fly proppant fracturing. In: Paper presented at International Petroleum Technology Conference. Beijing, China: International Petroleum Technology Conference. 2013. Paper Number: IPTC-16494-MS. DOI: 10.2523/IPTC-16494-MS

[113] Crews JB, Gomaa AM. Nanoparticle associated surfactant micellar fluids: An alternative to crosslinked polymer systems. In: Society of Petroleum

Engineers. Paper presented at SPE International Oilfield Nanotechnology Conference and Exhibition. Noordwijk, The Netherlands; 2012

[114] Al-Muntasheri GA. A critical review of hydraulic fracturing fluids for moderate- to ultralow-permeability formations over the last decade. SPE Production & Operation. 2014;**29**(4):243-260. DOI: 10.2118/169552-PA

[115] Clark JB. A hydraulic process for increasing the productivity of wells. Journal of Petroleum Technology. 1949;**1**(01):1-8. DOI: 10.2118/949001-G

[116] Perfetto R, Melo RCB, Martocchia F, Lorefice R, Ceccareli R, Tealdi L, et al. Oil-based fracturing fluid: First results in West Africa onshore. In: Paper presented at International Petroleum Technology Conference. Beijing, China; 2013

[117] Hutchins RD, Miller MJ. A circulating-foam loop for evaluating foam at conditions of use. SPE Production & Facilities. 2005;**20**(4): 286-294

[118] Karadkar P, Bataweel M, Bulekbay A, Alshaikh AA. Energized fluids for upstream production enhancement: A review. In: Paper presented at SPE Kingdom of Saudi Arabia Annual Technical Symposium and Exhibition. Dallas, Dammam, Saudi Arabia: Society of Petroleum Engineers (SPE); 2018

[119] Abdelaal A, Aljawad MS, Alyousef Z, Almajid MM. A review of foam-based fracturing fluids applications: From lab studies to field implementations. Journal of Natural Gas Science and Engineering. 2021;**95**:104236. DOI: 10.1016/j. jngse.2021.104236

[120] Wanniarachchi WAM, Ranjith PG, Perera MSA. Shale gas fracturing using

foam-based fracturing fluid: A review. Environmental Earth Sciences. 2017;**76**:91

[121] Harris PC, Heath SJ. High-quality foam fracturing fluids. In: Paper Presented at the SPE Gas Technology Symposium; Calgary, Alberta, Canada; April 1996. Paper Number: SPE-35600-MS. 1996. DOI: 10.2118/35600-MS

[122] Perdomo MEG, Madihi SW. Foam based fracturing fluid characterization for an optimized application in HPHT reservoir conditions. Fluids. 2022;7:156

[123] United States Environmental Protection Agency. Evaluation of Impacts to Underground Sources of Drinking Water by Hydraulic Fracturing of Coalbed Methane Reservoirs. Washington, DC: Office of Groundwater and Drinking Water; 2004

[124] Fink JK. Chapter 17-fracturing fluids. In: Petroleum Engineer's Guide to Oil Field Chemicals and Fluids. 1st ed. Waltham, Massachusetts: Gulf Professional Publishing, Elsevier; 2012

Chapter 2

A Review of Fracturing Technologies Utilized in Shale Gas Resources

Hisham Ben Mahmud, Mansur Ermila, Ziad Bennour
and Walid Mohamed Mahmud

Abstract

The modern hydraulic fracturing technique was implemented in the oil and gas industry in the 1940s. Since then, it has been used extensively as a method of stimulation in unconventional reservoirs in order to enhance hydrocarbon recovery. Advances in directional drilling technology in shale reservoirs allowed hydraulic fracturing to become an extensively common practice worldwide. Fracturing technology can be classified according to the type of the fracturing fluid with respect to the well orientation into vertical, inclined, or horizontal well fracturing. Depth, natural fractures, well completion technology, capacity, and formation sensitivity of a shale reservoir all play a role in the selection of fracturing fluid and fracturing orientation. At present, the most commonly used technologies are multi-section fracturing, hydra-jet fracturing, fracture network fracturing, re-fracturing, simultaneous fracturing, and CO_2 and N_2 fracturing. This chapter briefly reviews the technologies used in shale reservoir fracturing.

Keywords: hydraulic fracturing technology, unconventional reservoirs, fracturing fluids, well fracturing

1. Introduction

1.1 Development of fracturing technology

In the past four decades, various technologies have been developed and implemented to improve the production from shale gas formation as it is a commercially feasible source of energy. Hydraulic fracturing is a technique applied to enhance hydrocarbon extraction from subsurface geological formations by injecting a fluid at pressure higher than formation pressure to crack open the hydrocarbon formation rock. The hydraulic fracturing technology is not new; first experiment was conducted in 1947, and the first industrial implementation was in 1949 [1]. Hydraulic fracturing has, since then, been used for stimulating unconventional reservoirs and enhancing oil and natural gas recoveries. The first operation of fracturing treatment was performed by gelled crude, and later gelled kerosene was used. By the end of year 1952, many fracturing treatments were carried out by processed and live crude oils. This type of fluids is low-cost and permitting greater volumes at lower cost. In 1953 water-based fluids began to be utilized as a fracturing fluid, and a number of gelling agent additives such as surfactants were added, to the fracturing fluids, to reduce emulsion with formation fluid. Subsequently, additional clay stabilizing agents were improved and incorporated with water and used as a hydraulic fracturing fluid to fracture many reservoir formations. Alcohol and foam were also used to improve water-based fracturing fluids and utilized to fracture more formations. Currently aqueous fluids such as acid, brines, and water are utilized as base fluids with around 96% of all fracturing treatments using a propping agent. During the early years of the 1970s, the key advance in using fracturing fluids was in applying metal-based cross-linking agents to increase the viscosity of gelled water-based fracturing fluids designed for deeper wells at higher-temperature conditions [1].

The key factor of technological revolution is due to the fast evolution of drilling and completion techniques as well as the improvement of the fracturing technology. From the primary explosion technology of nitroglycerin to the newest fracturing technology of synchrotron, the developed fracturing technology has gradually improved the shale gas recovery efficiency.

The earliest nitroglycerin explosion technology was used in the 1970s in a vertical well with an open-hole completion. This technique affected wellbore stability and caused very limited penetrations. In 1981, a new fracturing fluid combined of nitrogen (N_2) and carbon dioxide (CO_2) foam was utilized in vertical wells in shale gas formations. This implementation led to gas recovery increase by 3–4 times and reduced formation damage. Subsequently, in 1992 the first horizontal well was drilled in shale gas formation in Hammett basin. Horizontal wells then

Stage	Year	Total well number	Fracturing technology
Initial	1979	5	High-energy gas fracturing
	1981	6	N_2 and CO_2 foam fracturing
	1984	17	Cross-linked gel fracturing, liquid quantity 105 gal (378 m3)
	1985	49	Cross-linked gel fracturing, liquid quantity 5 × 105 gal (1892 m3)
	1988	62	Cross-linked gel fracturing
	1991	96	Horizontal well and cross-linked gel fracturing
	1995	200	Horizontal well fracturing and cross-linked gel fracturing
	1997	300	Riverfracing treatment, liquid quantity 5 × 105 gal (1892 m3)
	1999	450	Riverfracing treatment, inclinometer fracture monitor
	2001	750	Riverfracing treatment, microseismic fracture monitor
	2002	1700	Horizontal well fracturing, riverfracing treatment
Development	2003	2600	New well configuration with 719 vertical wells, 85 horizontal wells, and 117 directional wells
	2004	3500	150 wells with horizontal well stage fracturing 2–4 stages
	2005	4500	600 new horizontal wells where drilling time is greatly reduced
	2006	5500	Synchronous fracturing, lower development costs
	2007	7000	Horizontal well fracturing, synchronous fracturing
	2008	9000	Repeated fracturing
Steady	2009	13,000	Maintain capacity, lower costs, enhancing oil recovery

Table 1.
Stimulation development of Barnett shale gas formation [3].

steadily supplanted the practice of vertical wells. A cross-linked gel was applied as a thickening or cross-linking agent during the period from the 1980s to the 1990s. The fracturing technique of horizontal wells can effectively generate fractured networks and increase the hydrocarbon flow area. This method is favorable because it minimizes the cost and increases hydrocarbon recovery. Thus, the development of large-scale hydraulic fracturing using horizontal wells contributed to the economic development of shale gas resources [2].

A major development was made in 1998 in fracturing technology by introducing a water-based liquid fluid instead of gel. This new fracturing fluid has a low sand (proppants) ratio of approximately 90% less than that used in the gelled fracturing. Thus, fracturing fluid associated cost was minimized by more than 50%. This type of fracture fluid can provide better fracturing performance that may increase the recovery efficiency up to 30% [2].

After the year 2000, a new technology called the segmental fracturing technology has been developed and utilized in horizontal wells during shale gas exploitation. This technology has further been developed and improved to include more than 20 segments leading to improvements in both the recovery efficiency and drainage area. Horizontal segmental fracturing technology is broadly used in the United States in the development of shale gas wells over the standard method by 85% [2].

After the year 2005 using both techniques of segmental fracturing technology and microseismic crack monitoring in shale gas development using fracture horizontal wells has significantly enhanced shale gas recovery. A new brand of fracturing technology was subsequently introduced in the year 2006 which is synchronous fracturing technology that has been utilized in the Barnett shale gas basin. **Table 1** summarizes the development of drilling and completion methods and the history of shale gas development in the Barnett basin, United States [3].

2. Main fracturing mechanisms of improving shale gas reservoir production

The mechanism of fracturing stimulation of shale gas reservoirs is not the same as a conventional or sandstone gas reservoir. Shale gas reservoirs, in general, cannot be

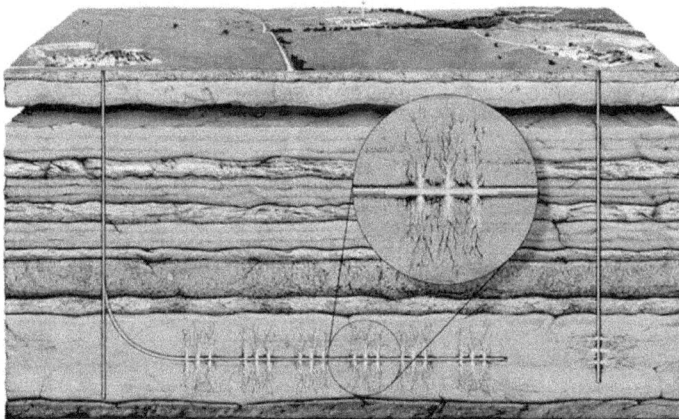

Figure 1.
Sketch map of vertical well and horizontal well fracturing [4].

found as conventional traps, but they are self-generating and self-storage gas reservoirs. The natural fracturing network can particularly enhance shale tight formation permeability [4]. Shale gas capacity can be attained through microfractures in shale formation. These fractures involve both a percolation path and a storage space of shale gas. They create the necessary communication and connectivity for the shale gas to reach the wellbore. Furthermore, shale gas recovery factor can be achieved through the existence of reservoir fractures' and its density and characteristic and

Fracturing technology	Technical physical features	Application area
Stage fracturing	• It is widely used with high technology maturity • Fracturing process conducting with multiple stages	• A horizontal well with multiple production zones and vertical stack tight reservoir
Riverfracing treatment	• Easy preparation of fracturing fluid with low cost • The main element of fracturing fluid is drag-reducing water, to create a denser fracture network, improving permeability • Forcing the gas to flow from the reservoir to the wellbore with greater ease • Less pollution impact on geological formation and limited sand carrying capacity	• Suitable to medium formation depth (1.5–3 km), • Natural fracture system developed reservoir
Hydra-jet fracturing	• Applied to create fractures at different directions and broaden the fracture network to increase hydrocarbon production • It does not require mechanical seal; thus it saves operational time	• Barefoot well completion
Repeated fracturing	• Reinstate the fracture to enhance fluid recovery. • Fracturing multiple wells simultaneously	• Development of new wells • Capacity decline of production well
Simultaneous fracturing	• It is a simultaneous operation process for multiple wells to save operation time • It has a better effect on the reservoir than fracture networks	• For reservoirs with big borehole density and nearby well location
Network fracturing	• Applying high-displacement fracturing fluid during the operation to open natural fracture and create network fractures • Increases formation permeability	• Low formation permeability in which natural fractures are not well developed
CO_2 and N_2 foam fracturing	• Causes less formation damage and pollution • Low filtration and good sand carrying capacity • Good for shale gas desorption	• Water-sensitive reservoir • Shallow reservoir (<1.5 km) and low well pressure
Large hydraulic fracturing	• Utilizes a huge amount of gel • High operation cost for well completion • Causes more damage to the reservoir	• No specific condition for the reservoir; thus it is widely used

Table 2.
Technical characteristics and application of fracturing technologies [7].

opening degree in the reservoir. Shale reservoirs are usually well stimulated and completed with good natural fractures and bedding. High brittleness is one of the significant parameters, which relates to the share failure during shale reservoir hydraulic fracturing process. It is responsible for the formation of complex fracture networks and the connections between natural fractures. Hence, the main purpose of utilizing stimulation technology on shale gas formation is to generate effective fracture networks to improve the reconstruction volume and enhance the reservoir capacity [5].

2.1 Main applied technology of shale reservoir fracturing

Fracturing technology of shale reservoirs can be classified based on the type of well fracturing into three categories, vertical, deviated, and horizontal fracturing wells, as shown in **Figure 1**. Fracturing technology can also be divided based on the type of fracturing fluid used such as gas, foam, gel, etc. Target zone can be fractured into different sections as single section and multi-section fracturing. Moreover, various factors should be taken into account while choosing the choice of fracturing fluid and fracturing technology such as the shale gas reservoir depth, capacity and formation sensitivity, natural fractures, and the well completion technology [6].

The most commonly used fracture technologies now are the multi-section fracturing, riverfracing, hydra-jet fracturing, fracture network fracturing, re-fracturing, and simultaneous fracturing. However, more attention is being given to CO_2 and N_2 fracturing. This fracturing technology's features and application conditions are different as shown in **Table 2**.

3. Multi-fracture network fracturing

Since it was proposed for the first time by Giger in 1985 [8], the concept of horizontal well fracturing has been widely practiced as a valuable technique to improve well production and increase the recovery of unconventional reservoirs. Horizontal well fracturing treatments in field generally create multi-fractures in selected intervals along the wellbore. Processes of fracture initiation and propagation in horizontal wells are different from those in vertical wells due to the larger contact surface area with the formations, thus resembling more complex reservoir situation. When multi-fractures are propagated, they often join or intersect with each other, forming patterns that are known as multi-fracture networks, which immensely increase the storage capacity and the fluid transmissibility of formations. Multi-fracture networks are not easy to be assessed or studied due to the complexity; however, they are evaluated using mathematical and statistical techniques and may be represented using fractals.

3.1 Creation of the multi-fracture network

The classical hydraulic fracturing theory indicates that the main formed fracture is a symmetric bi-wing plane extending parallel to the direction of maximum principal stress. However, field hydraulic fracturing treatment is completely different as complex fracture networks take place where the main fracture and other smaller branch fractures simultaneously extend in the fracture propagation zone [9–11].

Microseismic mapping shows that hydraulic fracturing in shale forms a multi-fracture network system [12–15] which consists of complex fractures as shown in **Figure 2** [16]. It was concluded from the mapping that natural fractures' direction was to the northwest and the propagation of the induced hydraulic fractures direction was to the northeast where they intersected with natural fractures. This led to many crosscutting linear features and formed a complex fracture. Based on fracture extension characteristic in shale reservoirs, hydraulic fractures are classified into four major types [16]: single plane bi-wing fracture, complex multiple

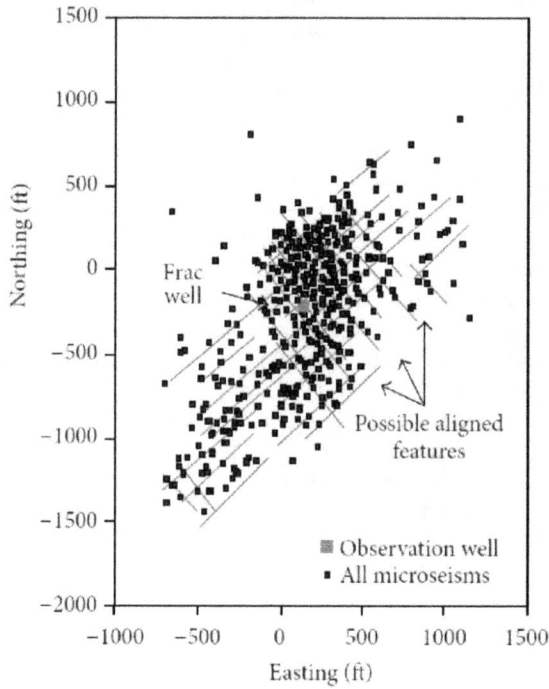

Figure 2.
Multi-fracture network extension in shale reservoirs during hydraulic fracturing (after Warpinski et al. 2008 [16]).

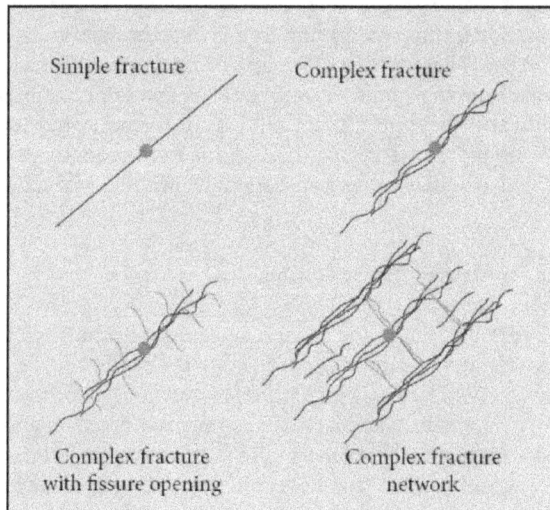

Figure 3.
The hydraulic fracture classification complexity (after Warpinski et al. 2008 [16]).

fracture, complex multiple fracture with open natural fractures, and complex fracture network as shown in **Figure 3**.

Confirming field observation from seismic mapping, simulation experiments [17–22] show that induced hydraulic fracture presents three types of extensions when intersecting with natural fractures: crossing the natural fractures, extending along the natural fractures or crossing, and extending along at the same time. It was concluded that fracture network would highly form during fracturing process of naturally fractured formations [23]. Moreover, several laboratory experiments confirmed that fracture network exists [24, 25] and found that the fracture network would easily form under low fluid viscosity injection [26, 27]. Other observations proposed that multi-fracture networks in shale reservoirs area are key to increase stimulated reservoir volume (SRV) where treatment success relies on whether hydraulic fracture could extend to form multi-fracture network [28–30].

3.2 Factors affecting multi-fracture network fracturing

Understanding fracture initiation and propagation rules are the main issues faced when commencing hydraulic fracturing because several important geological and engineering factors affecting the multi-fracture network formation are to be considered [31].

3.2.1 The geological factors.

1. Mineral composition. Brittleness is controlled by mineralogy as brittleness mineral concentration, the rock brittleness gets higher, and the development of natural fractures becomes better (mineral concentration increase/decrease).

2. Mechanical properties. Poisson's ratio and Young's modulus are combined to reflect the rock ability to fail under stress (Poisson's ratio) and maintain a fracture (Young's modulus) once the rock fractures. The lower Poisson's ratio and higher Young's modulus value, the more brittle the rock, and the fracture extends into fracture network.

3. Distribution of natural fractures. As natural fractures have great effect on hydraulic fracture extension, the more developed the natural fractures are, the more complex is the extension of hydraulic fracture.

4. Horizontal stress field. Multi-fracture network is controlled by intersecting intensity between induced fractures and natural fractures. Hydraulic fracture would propagate along natural fractures under low horizontal stress and cross natural fractures under high horizontal stress conditions.

3.2.2 The engineering factors.

1. Net fracturing pressure. Greater fracturing pressure would cause more complex fractures where it is possible to induce branches of hydraulic fracture to form a complex fracture network.

2. Fluid viscosity. The viscosity has an important influence on the complexity of fracture extension; from the laboratory experiments, it is obvious if the fluid viscosity gets higher; the complexity of fracture is significantly reduced. The injection of high viscosity fluid in field treating will reduce the complexity of fracture network [32–35].

3. Fracturing scale. The impact of fracturing scale can be seen on the production scale, as large amounts of the fracturing fluid volume are pumped; the longer the total length of fracture network, the more complex the resulted fracture network, and the higher the corresponding well production. Using large fracturing scale is an important measure to increase the SRV, which is essential to improve stimulation effect in the shale fracturing, where the bigger the SRV is, the higher the production.

The essential goal for the treatment is to get the most out of each stage and each cluster in the fracturing network. The optimization of fracturing fluid and minding the aforementioned factors can help achieving even flow distribution and network efficiency, both of which can help contribute to increased production. The practices over have realized that, in most cases where it has been measured, only 30–60% of the fractured clusters in a wellbore are providing measurable production [36].

4. Re-fracturing technology

Unconventional reservoirs show significant decline rates after few months of production compromising the economics and imposing the need for increasing or stabilizing production. The decline in production from the unconventional reservoirs is attributed to the closure and damage of the fracture networks within the formations. Hence, re-fracturing as an emerging technology has become a viable option for sustaining production and increasing reserves. Re-fracturing is a preferred option over drilling and completing new horizontal wells as it can be carried at only a fractional cost of up to 25–40% [37], thus minimizing the related financial and safety risks.

Production decline rates from unconventional reservoirs are more rapid than those in conventional reservoirs because of the ultralow permeability, limited reservoir contact, and the original completion strategy. The ability of re-fracturing technology provides a potential to extend the productive life of the unconventional reservoirs beyond the normal and up to an additional 20–30 years [38]. Re-fracturing restores production from underperforming formations by increasing fracturing networks, replacing damaged proppant, bypassing skin zones, and connecting old and new fractures [39]. Successful re-fracturing can increase the estimated ultimate recovery (EUR), shorten the capital return time, and increase the net present value (NPV) of the unconventional reservoirs. Decline curve analysis (DCA) showed that re-fractured wells achieved an average of 60% increase in NPV [40]; therefore, re-fracturing application helps reduce the variability in the unconventional reservoir performance and considered the best option for tackling production declines.

4.1 Re-fracturing process

Re-fracturing literally means a second hydraulic fracturing through same or new perforations to repair or recreate fracture networks within the same formation. If a re-fracturing treatment was carried out after a re-fracturing, then it would be considered a tri-fracturing [41].

Practically, re-fracturing is carried out when the initial hydraulic fracturing treatment was undersized or when suspected skin damage exists [42]. It is possible to use the existing fractures for the re-fracture and still generate a new fracture network sufficient to increase production. In a formation with its low in situ stress anisotropy, pressure can be created within the fracture itself to cause the reservoir to be fractured in new directions. Reusing the existed fractures helps control the cost of re-fracturing. Therefore, another approach for re-fracturing is to add perforations between the existing fractures to create additional fracturing networks as shown in **Figure 4**.

4.2 Re-fracturing methods

There are many ways available to perform re-fracturing; however, three most common re-fracturing methods are selected for consideration, namely, the diversion method, the coiled tubing fracturing method, and the mechanical isolation method [43]:

- Diversion: This method uses diverting agents to plug the existed fractures or perforations, allowing re-fracturing reallocation to new areas. However, it is difficult to control which segment of the lateral would be stimulated that is why it's also known as a "pump and pray method." Yet, this method is the most widely used in the industry likely because it is the most cost-effective.

- Coiled tubing: This method utilizes resettable packers where re-fracturing is targeted. However, at low rates through coiled tubing, this method is considered inadequate for open-hole environments.

- Mechanical isolation: This method typically uses expandable liners and plugs. However, it requires new hardware for re-fracturing which increase costs substantially because it would often need to use a full new liner.

As re-fracturing technology gains popularity in unconventional reservoirs, the ability to isolate reservoir access points and redirect the fracturing fluids and proppant to different parts of the reservoir is crucial to achieving a successful treatment. All known methods have advantages and disadvantages; however, the often selected method is based on their ease of use, cost-effectiveness, and environmental impact.

4.3 Selection for re-fracturing

Many wells are drilled with outdated completion designs; for that, they aren't efficiently producing the reservoir formations. These wells are specifically targeted when engaging re-fracturing because it is an economical practice to mitigate the flow rate decline and maximize reservoir deliverability [44].

The process of choosing which well to re-fracture is known as "candidate selection" [45], and the following are criteria which are often considered [46]:

- Logs or tracers indicating unproductive sections of wellbore

- Initial completion used wrong fracture fluid or proppant type

- Degree of production depletion

- Degradation in fracture conductivity or propped half-length

- Productivity of the reservoir

- Performance of other nearby wells

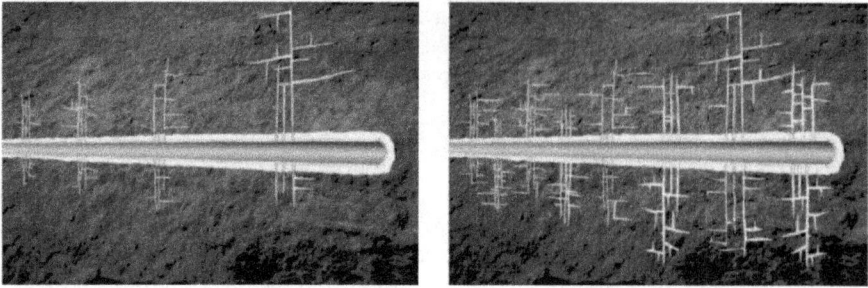

Figure 4.
(left) a hydraulic fracturing stimulation created a fracture network (right) after re-fracturing, and additional complex fracture network has developed (Allison & Parker 2014 [38]).

The selection methodology must be customized to fit the particular needs of a given field where substantial incremental reserves can be added if the correct candidate selection process is followed [47].

4.4 Evaluation of re-fracturing

After re-fracturing, a well may experience increase in production due to new fractures or extension of existing fracture networks. The success of re-fracturing can be determined by empirical parameters such as production rate 30 days before and following re-fracturing, EUR ratio based on DCA [48].

Computer programs can simulate re-fracturing scenarios at a considerable degree of accuracy despite the fact that all predictive methods lack robustness that accounts for the original production depletion and the conditions after re-fracturing. However, as technology advances, well performed computer models are able to generate trustworthy forecasts that allow decision-makers to confidently evaluate the economic success or failure of re-fracturing.

5. Simultaneous fracturing technology

Simultaneous fracturing or multiple fracturing (simul-frac) technology is the hydraulic fracturing technique that fractures multiple wells simultaneously. Simultaneous fracturing applies a shortest well-to-well distance to allow both the proppants and fracturing fluid flow through the porous medium from well to well under high pressure as shown in **Figure 5**. The purpose of the multiple simultaneous process is to increase the recovery efficiency and productivity, of the wells, by increasing the surface area subject to flow through the newly created dense fractures. The typical practice of simultaneous fracturing initiates with two horizontal wells of the same depth; however, currently up to four wells can be simultaneously fractured [46].

Many researchers have performed different field experiments to examine the simultaneous fracture multiple adjacent horizontal wells to create complex fracture networks. Even though field attempts have shown significant improvement with simul-frac instead of stand-along wells [50], microseismic information [51], and numerical simulations [52–58] also demonstrate a complex fracture network made through simul-frac. However, the reasons behind its success are not yet well understood. Multiple hydraulic fracture technique is a complex method that requires considering not only the hydraulic fracturing procedure but also fracture interaction between multiple fractures. The hydraulic fracturing treatment is a

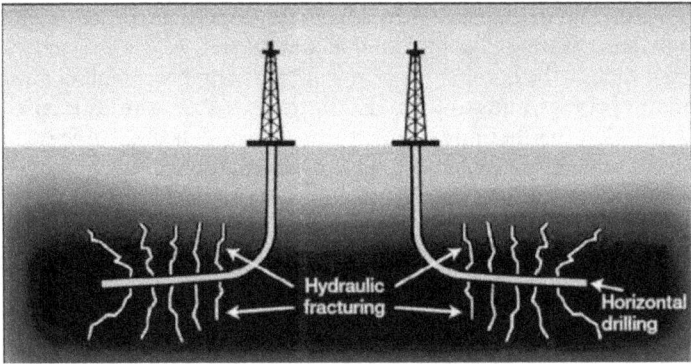

Figure 5.
An example of simultaneous fracturing [49].

typical hydromechanical fracture coupling problem, wherein the following three basic processes involve in [59]:

a. Rock deformation made by fluid pressure applied on fracture surface

b. Fluid flow into the fractures

c. Fracture growth

The fracture interaction between multiple fractures would significantly result in stress shadow effects that can cause stress field and fracture geometry alterations.

With the advance of computer processes, more numerical tools have been developed to become reliable and convenient techniques to investigate the treatment methods of hydraulic fracturing. Moreover, the numerical technique of finite element [60] is a well-established scheme to study rock engineering issues, and also it is frequently used in the last three decades to simulate hydraulic fracture propagation [61]. However, there are many scientific articles published on different finite element methods to numerically study the process of hydraulic fracturing [62–82].

6. Horizontal well staged fracturing technology

Horizontal well fracturing technology is the main technology promptly utilized to low permeability reservoirs. However, in deep shale reservoirs, the use of traditional single stimulation cannot meet the production requirements. Thus, a new technology of horizontal well pressure cracking has been introduced. Zebo et al. [83] found that, based on the process and concerned parameters of horizontal well fracturing, increasing technical problems during reservoir exploration and development, horizontal section becomes popular where sub-fractured horizontal well technique has wide application potentials. Furthermore, the sub-fracturing technology is an important tool in the technology of staged fracturing. Packer as a completion tool does not consist of multicolumn zones, and supporting tools are necessary for safety and to increase the possibility of successful fracturing treatment.

The success of horizontal well fracture is mainly due to the mechanical properties of the rock, stress, shaft stress fracture initiation, and elongation mechanism. Moreover, the horizontal well sub-fracturing should be considered to obtain better fracturing design and to ensure treatment success and efficiency. To achieve the expected outcomes from well completion of a fracturing job, certain issues must be monitored such as the borehole or near wellbore area, permeability anisotropy, blocking natural cracks, and stimulation failure. Up to date, the horizontal well fracturing technique has become one of the preferred tools to solve these problems. Thus, the main applied technology of horizontal well fracturing consists of limited flow fracturing technique and sub-fracturing process. The following section will describe these techniques.

6.1 Limiting entry fracturing

This technique limits the number of perforations and their diameter while injecting a large volume of fracturing fluid that causes increasing the bottom hole pressure on a large scale. Therefore, the fracturing fluid is forced to shunt into limited entries creating new fractures as shown in **Figure 6** [85, 86]. The main advantages of this technique are a relatively simple operation, short operation time, the fact that multi-fractures are created in a single operation which is environmentally favorable for reservoir protection. However, this technique has some limitations including high perforation back pressure, difficult to control any single fracture, and fractures which may not form in perforations of long interval horizontal well.

An example where limited entry fracturing technology was applied in horizontal well is Zhao 57-Ping 35 of Daqing Oil Field [84]. The well was divided into 4 sections each containing 19 perforations, and an isolating packer was set above the kickoff point. Using two simultaneous pumping facilities, a total fracturing fluid volume of $374.3m^3$ with an average sand ratio of 35.6% was injected at a rate of $7.5 \ m^3/min$. The fracture initiation pressure was 30.5 MPa, four fractures were created, and the total fracture span was 400 m. The entire operation took 79 minutes. This treatment achieved success allowing the production after fracturing to increase 20–30 times and reach the production level of 4 vertical wells.

6.2 Staged fracturing technique

As limited entry fracturing cannot operate on all the target layers at one time, staged fracturing technique is used when the horizontal section is long and many layers are targeted for fracturing. Staged fracturing creates many fractures by utilizing packers and/or other segmenting materials. Operating a section by section at the time, one fracture is created in every section. The key points to achieve staged fracturing are tools and technique that fulfill the treatment requirements.

There are three types of staged fracturing techniques often used: the bridge plug fracturing, through coiled tubing fracturing with straddle packer and gel

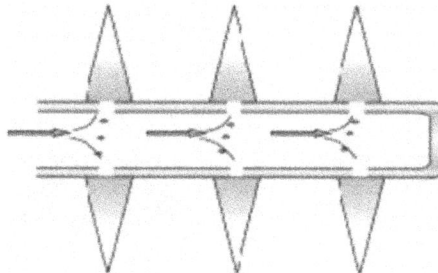

Figure 6.
Technique of the limited entry fracturing of a horizontal well [84].

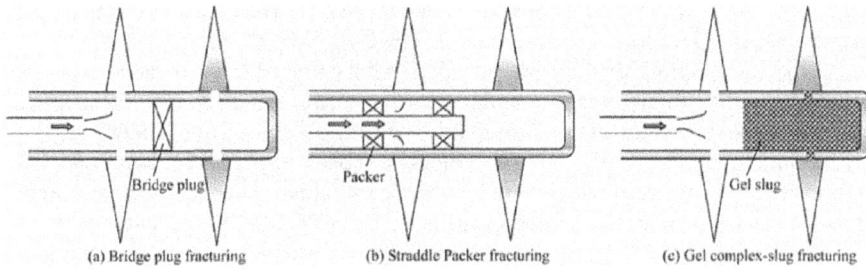

Figure 7.
Staged fracturing mechanism of horizontal wells [84].

complex-slug fracturing as shown in **Figure** 7. Contrary to packer separation, the gel complex-slug fracturing avoids the risk of downhole tool stuck, but in the latter, the fracture initiation points are difficult to control.

An example where gel staged fracturing technology was applied in well Saiping-1 of Changqing Oil Field where four fractures were created. The process is briefly described as the following: perforating the end of horizontal well section, followed by first fracturing treatment, running a production test, and temporary plugging the first section by sand filling gel plug and, next, repeating the process in perforating the second, third, and fourth sections followed by a formation pressure and production tests.

7. Evolution of fracturing fluid and the chemicals

The first hydraulic fracturing treatment was implemented in Hugoton Gas Field in Grand County, state of Kansas, during 1947. By the end of 1952, many fracturing treatments were performed with refined and crude oils. Thus oil-based fluids were the first fracturing fluid utilized for this purpose due to their benefits which are cheap and permitting greater volumes at a lower cost. But due to the safety and environmental issues, which are associated with their applications, it was encouraged that the industry move toward in developing an alternative fluid. At the beginning of 1953, for the first time, water fluid was used as a fracturing fluid; and a number of gelling agents were developed. However, water-based fluids with water-soluble polymers mixed to prepare a viscous solution are commonly used in the fracturing treatment. Since the late 1950s, more than 50% of the fracturing treatments were performed with fluids consisting of guar gums, high-molecular-weight polysaccharides composed of mannose and galactose sugars, or guar derivatives [87].

In 1964, surfactant agents were added to reduce the emulsion formation when in contact with the reservoir fluid; however, potassium chloride was added to decrease the effect on clays and other water-sensitive formation components. Later, additional clay stabilizing agents were developed to enhance the potassium chloride, allowing the use of water in different geological formations. In the early 1970s, a major revolution in fracturing fluids introduced the use of metal-based cross-linking agents to improve the viscosity of gelled water-based fracturing fluids for extreme reservoir condition (i.e., high temperature). Later a critical development was made on gelling agent to achieve a preferred viscosity. Also guar-based polymers are still used in fracturing jobs at reservoir temperatures below 150°C. Other fluid improvements, foams, and the addition of alcohol have enhanced the use of water in more geological reservoir formations. Moreover, various aqueous fluids, such as acid, gas, water, and brines, are currently used as the base fluid in approximately 96% of all fracturing treatments employing a propping agent [87].

As the hydrocarbon drilling and production have moved toward deeper reservoirs with high pressure and temperature condition, more fracturing treatments have been developed to be compatible with these conditions. Therefore, gel stabilizers and thermally stable polymers have been developed in which gel stabilizers can be utilized with around 5% methanol, but synthetic polymers have shown a sufficient viscosity at temperatures up to 230°C [88]. After that, chemical stabilizers have been developed and possibly used with or without a methanol. The improvements, which are made in cross-linkers and gelling agents, have led to systems that can permit the fluid to reach the well bottomhole in high-temperature condition before cross-linking, therefore, reducing the effects of high shear in the production tubing. Recently, nanotechnology has been introduced in the design of new, efficient hydraulic fracturing fluids [88]. For example, nanolatex silica is used to reduce the concentration of boron found in conventional cross-linkers. Recent advancement in nanotechnology is the use of small-sized silica particles [20 nm] suspended in guar gels to improve fracturing treatment [89]. Therefore, the following section will discuss the use of CO_2 and N_2 as fracturing fluid to enhance the hydrocarbon fluid production and to store CO_2 into the geological formation to minimize the greenhouse emission. Also it will provide a brief information on hydra-jet fracturing.

7.1 Fracturing using CO_2 and N_2

In the ordinary fracturing, large amounts of freshwater, sand, and chemicals are injected into the ground at high pressure. It has been reported that up to 9.6 million gallons of water on average are used for a single well fracturing; this lead to the use of more than 28 times the water for wells before fracturing, putting farming, and drinking sources at risk in arid regions, especially during drought [90]. Some of the water used for fracking is brought back to the surface and recycled, but the most of it is lost deep into the formations. Thus, fracking can increase demand for water by up to 30 percent, and this can be a major increase for groundwater consumption.

To solve the water scarcity problem, the fracturing using water, carbon dioxide, and nitrogen is commonly referred to the process in where substantial quantities of both nitrogen and carbon dioxide are incorporated into the fracturing fluid. Amounts of nitrogen and carbon dioxide are incorporated separately into an aqueous-based fracturing fluid to provide a volume ratio of nitrogen to carbon dioxide within an estimated range between 0.2 and 1.0 at wellhead conditions. The volume ratio for the total of both carbon dioxide and nitrogen to the aqueous phase of the aqueous fracturing fluid ranges between 1 and 4. The aqueous fracturing fluid that contains the nitrogen and carbon dioxide is injected in the well under conditions in which the pressure required is high enough to implement hydraulic fracturing of the subterranean formation undergoing treatment. In order to provide a viscous aqueous-based fracturing fluid, a thickening agent may be added into water. Additionally, a propping agent is to be incorporated into a portion of the fracturing fluid. Only then can carbon dioxide and nitrogen be added to the fluid. Carbon dioxide is incorporated in its liquid phase and the nitrogen in its gaseous phase. The use of carbon dioxide and nitrogen as fracturing fluids is discussed briefly in this essay.

Currently, carbon dioxide fracturing is one of the most effective and cleanest approaches available in order to increase oil and gas production. To produce the viscous aqueous-based fracturing fluid, carbon dioxide is injected in its liquid state using conventional frac pumps. Injection rates for it can be improved by incorporating booster capacity. An upside of using carbon dioxide in this process is that it can carry high concentrations of proppant in foam form due to its density and is compatible with all treating fluids (including acids). Because of that density, it is also not susceptible to gravity separation. Additionally, carbon dioxide can be pumped with synthetic and natural polymers, lease crude, or diesel as a foam or microemulsion, increasing the hydrostatic head to or greater than that of fresh water and decreasing the viscosity of the system. This feature of carbon dioxide results in vastly reducing horsepower costs and a decrease in the applied treating pressures. Another benefit of carbon dioxide is that it dissolves in water which causes it to form carbonic acid that dissolves the matrix in carbonate rocks. It buffers water-based systems to a pH of 3.2 which can also control clay swelling and iron and aluminum hydroxide precipitation. Known to act as a surfactant to significantly reduce interfacial tension and resultant capillary forces, carbon dioxide thus removes fracturing fluid, connate water, and emulsion blocks. In regard to it being one of the cleanest approaches in increasing gas and oil productions, carbon dioxide provides the energy to remove formations fines, crushed proppant, reaction products, and mud that is lost during drilling. In addition to that, swabbing of treating fluids can be greatly reduced which will allow for saving in associated treatment costs. Lastly, unlike other agents a carbon dioxide treatment with a 70 quality foam job allows low amounts of the water to contact the formation, roughly 30 percent compared to a gelled water fracturing. This decrease chances of clay swelling and inhibited production. All these benefits of using carbon dioxide as a fracturing fluid in wells with low bottomhole pressure or sensitivity to certain fluids make it a strong alternative candidate.

Although containing different properties, nitrogen similar to carbon dioxide comes with many benefits for fracturing fluids. Nitrogen for the fracturing fluids can be supplied by air products and provides both performance and cost advantages over certain formations of water-based fluids. Although water-based fracturing fluids are commonly used for hydraulic fracturing due to their advanced proppant transport into the fracture, they do also come with disadvantages. Because they can cause water saturation around the fracture and clay swelling which can result in hindering the mass transport of hydrocarbons from the fracture to the wellbore, water-based fluids are often unsuitable for water-sensitive formations. Nitrogen fracking fluids are an excellent alternative to water-based fluids in water-sensitive formations, depleted reservoirs, and shallow formations as they do not result in any water saturation.

Four main types of nitrogen fracturing fluids are used commercially: pure gas, foam, energized, and ultrahigh quality (mists). Foam fracturing fluids typically consist of a water-based system and a gas phase of nitrogen volume in the range of 53 to 95%. Below 53% nitrogen, the fracturing fluid is considered energized. Above 95 percent nitrogen, the fracturing fluid is considered a mist. Cryogenic liquid nitrogen fracking fluid is considered to be the fifth type of nitrogen fracturing fluids used. However, it is rarely employed for commercial operations due to material restrictions and equipment requirements.

7.2 Hydra-jet fracturing

The process of hydra-jet fracturing combines hydra-jetting with hydraulic fracturing and involves running a specialized jetting tool on conventional or coiled tubing. Dynamic fluid energy jets form tunnels in the reservoir rock at precise locations to initiate the hydraulic fracture which is then extended from that point outwards. By repeating the process, one can create multiple hydraulic fractures along the horizontal wellbore [91–93]. The idea of hydra-jet fracturing is not a new one.

In fact, it was used a century ago with low-pressure jets [94] where waterjets with erosive materials were used to cut rock and glass. Because erosion does not involve a backflow hindering the sand cutting process, cutting steel plates, wellheads during the Iraqi war, and rock quarries tend to be easily be done. Hydra-jet cutting may be mistakenly claimed as a result of a perforating process which can be seen when used on the rocks sandstone and limestone.

For these two rocks, assume that the jet is used to perforate formation rock. Also assume that the jetting process creates a perforation with a larger inside diameter than the jet nozzle. The velocity of the fluid flowing into the perforation tunnel would be incredibly elevated. Near the bottom of the perforation, the velocity of the flowing fluid would dramatically decrease. If the flow area is sustained and there is no presence of friction, the fluid pressure will be equal to the original jet pressure per the example. However, this tends to be an unlikely happening because pressure losses are typically high. To further explain this, jet boundary friction works to convert kinetic energy to heat loss causing jet flaring. This drastically reduces jet velocity, which in turn reduces the pressure per unit area of impact. This results in a low-pressure transformation efficiency. More importantly, rocks can still be fractured when enough pressure is applied to the jets even at this low of a pressure efficiency rate. An important note is that laboratory tests have shown that rock fracturing is commonplace when jet pressures are high. However, when high-pressure and low-energy transformation efficiencies are used hand in hand, they are technically and economically impractical.

8. Summary

The desired objective of fracturing is to develop and effectively produce from a shale reservoir. To ensure a successful fracturing treatment, a proper fracturing technology must be utilized based on the reservoir characteristics as the reservoir mineral content, physical properties, and geological condition. The utilized formation fracturing technique has a different desired environment to achieve the maximal recovery. During the process of fracturing treatment, the content of a fracturing fluid should be checked based on the formation mineral content and physical properties to improve reservoir permeability and reduce formation damage.

The forming of multi-fracture network is the key to obtain an effective hydraulic fracturing treatment in shale reservoirs. If higher treating net pressure is achieved, lower fluid viscosity is used, and larger fracturing scale attempt would be more helpful to form a fully fracture network. The reservoir geological factors also have high attributes, where brittleness index, elastic characteristic of rock mechanical properties, horizontal stress, and existence of natural fractures are useful to obtain better results of fractures developing into multi-fracture network.

Re-fracturing has the potential to re-energize natural fractures and extend and replace low conductivity existing fracture network. Utilizing re-fracture treatment successfully depends on technology that allows access to larger volumes of unconventional reservoirs. Monitoring the effectiveness of well completions helps guide technologies and methods to gain control of the wellbore to maximize EUR and NPV. Re-fracturing treatments have significant impact on production, and economics of unconventional reservoir development and consideration should be taken to determine the best way to achieve successful re-fracturing as production starts to decline.

A Review of Fracturing Technologies Utilized in Shale Gas Resources
DOI: http://dx.doi.org/10.5772/intechopen.92366

Author details

Hisham Ben Mahmud[1*], Mansur Ermila[2], Ziad Bennour[1]
and Walid Mohamed Mahmud[3]

1 Curtin University Malaysia, Sarawak, Malaysia

2 Colorado School of Mines, Colorado, USA

3 University of Tripoli, Tripoli, Libya

*Address all correspondence to: hisham@curtin.edu.my

IntechOpen

References

[1] Gandossi L, Von Estorff U. An overview of hydraulic fracturing and other formation stimulation technologies for shale gas production—Update 2015; EUR 26347. 2015. DOI: 10.2790/379646

[2] Wang Q. Influence of reservoir geological characteristics on fracturing fluid flowback [Master's thesis]. Calgary, AB: University of Calgary; 2017. DOI: 10.11575/PRISM/26519. (Unpublished)

[3] Jaripatke O, Grieser B, Chong KK. A Completions Road Map to Shale Play Development—Review of Successful Approach Towards Shale Play Stimulation in The Last Two Decades. SPE130369. United States of America: Society of Petroleum Engineers (SPE); 2010

[4] Ma C, Huang L, et al. Gas fracturing technique for shale and its effect evaluation. Tuha Oil Gas. 2011;**16**(3):243-246

[5] Zhao J, Wang S, et al. Difficulties and technical keys of fracturing reformation in shale gas reservoir. Natural Gas Industry. 2012;**32**(4):46-49

[6] Zhang R, Li G, et al. Present situation and prospect of shale gas production increasing technology. Petroleum Machinery. 2011;**39**(Suppl):117-120

[7] Zhang H, Yang Y, et al. Shale gas fracturing technology. Xinjiang Petroleum Science & Technology. 2013;**23**(2):31-35

[8] Giger FM. Horizontal Wells Production Techniques in Heterogeneous Reservoirs. SPE 13710. United States of America: Society of Petroleum Engineers (SPE); 1985

[9] Warpinski NR, Teufel LW. Influence of geologic discontinuities on hydraulic fracture propagation. Journal of Petroleum Technology. 1987;**39**(2):209-220

[10] Warpinski NR. Hydraulic fracturing in tight, fissured media. Journal of Petroleum Technology. 1991;**43**(2):146-151

[11] Warpinski NR, Lorenz JC, Branagan PT, Myal FR, Gall BL. Examination of a cored hydraulic fracture in a deep gas well. SPE Production & Facilities. 1993;**8**(3): 150-158

[12] Fisher MK, Wright CA, Davidson BM. Integrating fracture mapping technologies to optimize stimulations in the Barnett shale. In: Proceedings of the SPE Annual Technical Conference and Exhibition. United States of America: Society of Petroleum Engineers (SPE); 2002. pp. 975-981

[13] Fisher MK, Heinze JR, Harris CD, Davidson BM, Wright CA, Dunn KP. Optimizing horizontal completion techniques in the Barnett shale using microseismic fracture mapping. In: Proceedings of the SPE Annual Technical Conference and Exhibition. United States of America: Society of Petroleum Engineers (SPE); 2004

[14] Maxwell SC, Urbancic TI, Steinsberger N, Zinno R. Microseismic imaging of hydraulic fracture complexity in the Barnett shale. In: Proceedings of the SPE Annual Technical Conference and Exhibition. United States of America: Society of Petroleum Engineers (SPE); 2002. pp. 965-973

[15] Urbancic TI, Maxwell SC. Microseismic imaging of fracture behavior in naturally fractured reservoirs. In: Proceedings of the SPE/ISRM Rock Mechanics Conference. United States of America: Society of

Petroleum Engineers (SPE), International Society for Rock Mechanics (ISRM); 2002

[16] Warpinski NR, Mayerhofer MJ, Vincent MC, Cipolla CL, Lolon EP. Stimulating unconventional reservoirs: Maximizing network growth while optimizing fracture conductivity. In: Proceedings of the SPE Unconventional Reservoirs Conference. United States of America: Society of Petroleum Engineers (SPE); 2008. pp. 237-255

[17] Blanton TL. An experimental study of interaction between hydraulically induced and pre-existing fractures. In: Proceedings of the SPE Unconventional Gas Recovery Symposium. Pennsylvania, USA; 1982

[18] Blanton TL. Propagation of hydraulically and dynamically induced fractures in naturally fractured reservoirs. In: Proceedings of the SPE Unconventional Gas Technology Symposium. United States of America: Society of Petroleum Engineers (SPE); 1986

[19] Chen M, Pang F, Jin Y. Experiments and analysis on hydraulic fracturing by a large-size triaxial simulator. Chinese Journal of Rock Mechanics and Engineering. 2000;**19**:868-872

[20] Zhou J, Chen M, Jin Y, Zhang G. Experimental study on propagation mechanism of hydraulic fracture in naturally fractured reservoir. Acta Petrolei Sinica. 2007;**28**(5):109-113

[21] Zhou J, Chen M, Jin Y, Zhang G. Experiment of propagation mechanism of hydraulic fracture in multi-fracture reservoir. Journal of China University of Petroleum. 2008;**32**(4):51-54

[22] Chen M, Zhou J, Jin Y, Zhang G. Experimental study on fracturing features in naturally fractured reservoir. Acta Petrolei Sinica. 2008;**29**(3):431-434

[23] Mahrer KD. A review and perspective on far-field hydraulic fracture geometry studies. Journal of Petroleum Science and Engineering. 1999;**24**(1):13-28

[24] Beugelsdijk LJL, De Pater CJ, Sato K. Experimental hydraulic fracture propagation in a multi-fractured medium. In: Proceedings of the SPE Asia Pacific Conference on Integrated Modelling for Asset Management. United States of America: Society of Petroleum Engineers (SPE); 2000. pp. 177-184

[25] Bennour Z, Ishida T, Nagaya Y, et al. Fracture development and mechanism in shale cores by viscous oil, water and L-CO 2 injection. In: 48th US Rock Mechanics/Geomechanics Symposium, ARMA-2014-7164. Minnesota, USA; 2014

[26] Ishida T, Chen Y, Bennour Z, et al. Features of CO2 fracturing deduced from acoustic emission and microscopy in laboratory experiments. Journal of Geophysical Research, Solid Earth. 2016;**121**:8080-8098. DOI: 10.1002/2016JB013365

[27] Bennour Z, Ishida T, Nagaya Y, et al. Crack extension in hydraulic fracturing of shale cores using viscous oil, water, and liquid carbon dioxide. Rock Mechanics and Rock Engineering. 2015;**48**:1463. DOI: 10.1007/s00603-015-0774-2

[28] Mayerhofer MJ, Lolon EP, Youngblood JE, Heinze JR. Integration of microseismic fracture mapping results with numerical fracture network production modeling in the Barnett shale. In: Proceedings of the SPE Annual Technical Conference and Exhibition. United States of America: Society of Petroleum Engineers (SPE); 2006

[29] Mayerhofer MJ, Lolon EP, Warpinski NR, et al. What is stimulated reservoir volume (SRV)? In: Proceedings of the SPE Production &

Operations. United States of America: Society of Petroleum Engineers (SPE); 2010. pp. 89-98

[30] Bennour Z, Watanabe S, Chen Y, et al. Evaluation of stimulated reservoir volume in laboratory hydraulic fracturing with oil, water and liquid carbon dioxide under microscopy using the fluorescence method. Geo-Mechanics and Geo-Physics for Geo-Energy and Geo-Resources. 2018;**4**:39. DOI: 10.1007/s40948-017-0073-3

[31] Ren L, Zhao J, Hu Y. Hydraulic fracture extending into network in shale: Reviewing influence factors and their mechanism. The Scientific World Journal. 2014;**2014**:9. DOI: 10.1155/2014/847107

[32] Cipolla CL, Jensen L, Ginty W, De Pater CJ. Complex hydraulic fracture behavior in horizontal wells, south Arne field, Danish North Sea. In: Proceedings of the SPE Annual Technical Conference and Exhibition. United States of America: Society of Petroleum Engineers (SPE); 2000. pp. 35-47

[33] Cipolla CL, Hansen KK, Ginty WR. Fracture treatment design and execution in low-porosity chalk reservoirs. SPE Production & Operations. 2007;**22**(1):94-106

[34] Warpinski NR, Kramm RC, Heinze JR, Waltman CK. Comparison of single- and dual-array microseismic mapping techniques in the Barnett shale. In: Proceedings of the SPE Annual Technical Conference and Exhibition. United States of America: Society of Petroleum Engineers (SPE); 2005. pp. 913-922

[35] Cipolla CL, Lolon EP, Dzubin B. Evaluating stimulation effectiveness in unconventional gas reservoirs. In: Proceedings of the SPE Annual Technical Conference and Exhibition. United States of America: Society of Petroleum Engineers (SPE); 2009. pp. 3397-3417

[36] Walzel B. "Hydraulic Fracturing: Locking in Efficiencies" Operational Flexibility and Efficiencies Drive Hydraulic Fracturing Innovations. Houston, United States America: Hart Energy Publications; 2019

[37] Asala HI, Ahmadi M, Taleghani A. Why re-fracturing works and under what conditions? In: Proceedings of SPE Annual Technical Conference and Exhibition. Dubai, UAE; 2016

[38] Allison D, Parker M. Re-fracturing extends lives of unconventional reservoirs. In: The American Oil and Gas Reporter. Tech Trends. Derby, KS, United States America: The Better Business Publication; 2014

[39] Jacobs T. Changing the equation: Re-fracturing shale oil wells. Journal of Petroleum Technology. SPE-0415-0044-JPT. 2015;**67**(4):44-49

[40] Oruganti Y, Mittal R, McBurney C, Alberto R. Re-fracturing in Eagle Ford and Bakken to increase reserves and generate incremental NPV: Field study. In: Proceeding of SPE Hydraulic Fracturing Technology Conference. Texas, USA; 2015

[41] Wolhart S, McIntosh G, Zoll M, Weijers L. Surface tiltmeter mapping shows hydraulic fracture reorientation in the Codell Formation, Wattenberg Field, Colorado. In: Proceedings of SPE Annual Technical Conference and Exhibition. Anaheim, California, USA; 2007

[42] Santos L, Taleghani A, Li G. Smart expandable proppants to achieve sustainable hydraulic fracturing treatments. In: Proceedings of SPE Annual Technical Conference & Exhibition. SPE-181391-MS. Dubai, UAE; 2016

[43] Markit IHS. The emerging technology of re-fracturing horizontal wells. Energy & Natural Resources. London: IHS Markit Energy Expert; 2017. (retrieved)

[44] Yang C, Xue X, Huang J, Datta-Gupta A, King M. Rapid refracturing candidate selection in shale reservoirs using drainage volume and instantaneous recovery ratio. In: Unconventional Resources Technology Conference. San Antonio, Texas; 2016. pp. 1-3

[45] Moore L, Ramakrishnan H. Restimulation: Candidate selection methodologies and treatment optimization. In: Adapted from AAPG Annual Convention, San Antonio, Texas, April 20-23. Data and Consulting Services, Schlumberger. 2008

[46] Chen S, Du L, et al. Study on multi well simultaneous volume fracturing technology. Oil Drilling & Production Technology. 2011;**33**(6):59-65

[47] Crowell R, Jennings A. A diagnostic technique for restimulation candidate selection. In: SPE Annual Fall Technical Conference and Exhibition, 1-3 October. Houston, Texas; 1978. DOI: 10.2118/7556-MS

[48] Loyd E, Grieser W, McDaniel BW, Johnson B, Jackson R, Fisher K. Successful Application of Hydrajet Fracturing on Horizontal Wells Completed in a Thick Shale Reservoir. SPE-91435-MS. United States of America: Society of Petroleum Engineers (SPE); 2004. DOI: 10.2118/91435-MS

[49] Muresan JD, Ivan MV. Controversies regarding costs, uncertainties and benefits specific to shale gas development. Sustainability. 2015;7:2473-2489

[50] Mutalik PN, Gibson B. Case history of sequential and simultaneous fracturing of the Barnett shale in Parker county. In: Proceedings of the SPE Annual Technical Conference and Exhibition, ATCE, 2008. 2008. pp. 3203-3209

[51] Waters G, Dean B, Downie R, Kerrihard K, Austbo L, McPherson B.

Simultaneous hydraulic fracturing of adjacent horizontal wells in the Woodford shale. In: Proceedings of the SPE Hydraulic Fracturing Technology Conference. United States of America: Society of Petroleum Engineers (SPE); 2009;**2009**:694-715

[52] Olson JE. Multi-fracture propagation modeling: Applications to hydraulic fracturing in shales and tight gas sands. In: Proceedings of the 42nd U.S. Rock Mechanics - 2nd U.S.-Canada Rock Mechanics Symposium 2008. United States of America; 2008

[53] Yongtao Y, Xuhai T, Hong Z, Quansheng L, Zhijun L. Hydraulic fracturing modeling using the enriched numerical manifold method. Applied Mathematical Modelling. 2018;**53**:462-486

[54] Dahi-Taleghani A, Olson JE. Numerical modeling of multistranded-hydraulic-fracture propagation: Accounting for the interaction between induced and natural fractures. SPE Journal. 2011;**16**(3):575-581

[55] Weng X, Kresse O, Cohen C, Wu R, Gu H. Modeling of hydraulic-fracture-network propagation in a naturally fractured formation. SPE Production and Operations. 2011;**26**(4):368-380

[56] Wu R, Kresse O, Weng X, Cohen C, Gu H. Modeling of interaction of hydraulic fractures in complex fracture networks. In: Proceedings of the SPE Hydraulic Fracturing Technology Conference. The Woodlands, Texas, USA; 2012

[57] Yongtao Y, Xuhai T, Hong Z, Quansheng L, Lei H. Tree-dimensional fracture propagation with numerical manifold method. Engineering Analysis with Boundary Elements. 2016;**72**:65-77

[58] Nagel NB, Sanchez-Nagel M. Stress shadowing and microseismic events:

a numerical evaluation. United States of America: Society of Petroleum Engineers (SPE); 2011

[59] Adachi J, Siebrits E, Peirce A, Desroches J. Computer simulation of hydraulic fractures. International Journal of Rock Mechanics and Mining Sciences. 2007;**44**(5):739-757

[60] Zienkiewicz OC, Taylor RL. The Finite Element Method. London: McGraw-Hill; 2008

[61] Sato K, Itaoka M, Hashida T. FEM simulation of mixed mode crack propagation induced by hydraulic fracturing. In: Processing and Properties of Porous Nickel Titanium. United Kingdom: Hindawi Limited; 2013

[62] Paluszny A, Zimmerman RW. Numerical simulation of multiple 3D fracture propagation using arbitrary meshes. Computer Methods Applied Mechanics and Engineering. 2011;**200**(9):953-966

[63] Paluszny A, Tang X, Nejati M, Zimmerman RW. A direct fragmentation method with Weibull function distribution of sizes based on fnite- and discrete element simulations. International Journal of Solids and Structures. 2016;**80**:38-51

[64] Moes N, Dolbow J, Belytschko T. A finite element method for crack growth without remeshing. International Journal for Numerical Methods in Engineering. 1999;**46**(1):131-150

[65] Gordeliy E, Peirce A. Coupling schemes for modeling hydraulic fracture propagation using the XFEM. Computer Methods Applied Mechanics and Engineering. 2013;**253**(1):305-322

[66] Lecampion B. An extended finite element method for hydraulic fracture problems. Communications in Numerical Methods in Engineering. 2009;**25**(2):121-133

[67] Su K, Zhou X, Tang X, Xu X, Liu Q. Mechanism of cracking in dams using a hybrid FE-meshfree method. International Journal of Geomechanics. 2017;**17**(9):04017071

[68] Gupta P, Duarte CA. Simulation of non-planar three-dimensional hydraulic fracture propagation. International Journal for Numerical and Analytical Methods in Geomechanics. 2014;**38**(13):1397-1430

[69] Cruse TA. Numerical solutions in three dimensional elastostatics. International Journal of Solids and Structures. 1969;**5**(12):1259-1274

[70] Zhou D, Zheng P, He P, Peng J. Hydraulic fracture propagation direction during volume fracturing in unconventional reservoirs. Journal of Petroleum Science and Engineering. 2016;**141**:82-89

[71] Behnia M, Goshtasbi K, Zhang G, Yazdi SHM. Numerical modeling of hydraulic fracture propagation and reorientation. European Journal of Environmental and Civil Engineering. 2015;**19**(2):152-167

[72] Mukhopadhyay NK, Maiti SK, Kakodkar A. A review of SIF evaluation and modelling of singularities in BEM. Computational Mechanics. 2000;**25**(4):358-375

[73] Wittel FK, Carmona HA, Kun F, Herrmann HJ. Mechanisms in impact fragmentation. International Journal of Fracture. 2008;**154**(1-2):105-117

[74] Damjanac B, Gil I, Pierce M, Sanchez M, As AV, Mclennan J. A new approach to hydraulic fracturing modeling in naturally fractured reservoirs. Technology & Health Care. 2010;**18**(4-5):325-334

[75] Marina S, Derek I, Mohamed P, Yong S, Imo-Imo EK. Simulation of the hydraulic fracturing process of fractured rocks by the discrete element

method. Environmental Earth Sciences. 2015;**73**(12):8451-8469

[76] Munjiza A. The Combined Finite-Discrete Element Method. Hoboken, New Jersey, United States of America: John Wiley & Sons; 2004

[77] Mahabadi OK, Randall NX, Zong Z, Grasselli G. A novel approach for micro-scale characterization and modeling of geomaterials incorporating actual material heterogeneity. Research Letters. 2012;**39**(1):1303

[78] Lei Q, Latham J-P, Xiang J. Implementation of an empirical joint constitutive model into finite-discrete element analysis of the Geomechanical behaviour of fractured rocks. Rock Mechanics and Rock Engineering. 2016;**49**(12):4799-4816

[79] Latham JP, Xiang J, Belayneh M, Nick HM, Tsang CF, Blunt MJ. Modelling stress-dependent permeability in fractured rock including effects of propagating and bending fractures. International Journal of Rock Mechanics and Mining Sciences. 2013;**57**:100-112

[80] Lei Q, Latham JP, Xiang J, Lang P. Coupled FEMDEM-DFN model for characterising the stress-dependent permeability of an anisotropic fracture system. In: International Conference on Discrete Fracture Network Engineering. Vancouver, Canada; 2014

[81] Obeysekara A, Lei Q, Salinas P, et al. A fluid-solid coupled approach for numerical modeling of near-wellbore hydraulic fracturing and flow dynamics with adaptive mesh refinement. In: Proceedings of the 50th US Rock Mechanics/Geomechanics Symposium 2016. United States of America; 2016. pp. 1688-1699

[82] Yan C, Zheng H, Sun G, Ge X. Combined finite-discrete element method for simulation of hydraulic fracturing. Rock Mechanics and Rock Engineering. 2016;**49**(4):1389-1410

[83] Zebo L, Dawei L, Yang L. The research & development of horizontal well fracturing technology. Physical and Numerical Simulation of Geotechnical Engineering. 2013;(13):17-20

[84] Wenbin C, Zhaomin L, Xialin Z, Bo Z, Qi Z. Horizontal well fracturing technology for reservoirs with low permeability. Petroleum Exploration and Development. 2009;**36**(1):80-85

[85] Aishan L, Yang B, Yuqin J, et al. Viscoelastic surfactant fracturing fluid rheology. Petroleum Exploration and Development. 2007;**34**(1):89-92

[86] Tingxue J, Yiming Z, Xingkai F, et al. Hydraulic fracturing technology in clay-carbonate fractured reservoirs with high temperature and deep well depth. Petroleum Exploration and Development. 2007;**34**(3):348-353

[87] Montgomery CT, Smith MB. Hydraulic fracturing: History of An Enduring Technology, JPT. United States of America: Society of Petroleum Engineers (SPE). 2010. Available from: https://www.ourenergypolicy.org/wp-content/uploads/2013/07/Hydraulic.pdf

[88] Al-Muntasheri GA. A critical review of hydraulic-fracturing fluids for moderate- to ultralow-permeability formations over the last decade. SPE Production & Operations. 2014;**29**(4):243-260. DOI: 10.2118/169552-PA

[89] Shah SN, Fakoya MF. Rheological properties of surfactant-based and polymeric nano-fluids. In: SPE/ICoTA Coiled Tubing & Well Intervention Conference & Exhibition. United States of America: Society of Petroleum Engineers (SPE), Intervention and Coiled Tubing Association (ICoTA); 2013. DOI: 10.2118/163921-ms

[90] Magill B. Study: Water use skyrockets as fracking expands.

Researching and Reporting the Science
and Impacts of Climate Change. New
Jersey, United States of America:
Climate Central: A Science & News
Organization; 2015. Copyright © 2020
Climate Central

[91] McDaniel B, Surjaatmadja JB.
Hydrajetting Applications in Horizontal
Completions to Improve Hydraulic
Fracturing Stimulations and
Improve ROI. SPE-125944-MS2009.
United States of America: Society
of Petroleum Engineers (SPE).
DOI: 10.2118/125944-MS

[92] Gokdemir OM, Liu Y, Qu H,
Cheng K, Cheng Z. New Technique:
Multistage Hydra-jet Fracturing
Technology for Effective Stimulation
on the First U-Shape Well in Chinese
Coal Bed Methane and Case Study.
OTC-23987-MS. USA: Offshore
Technology Conference; 2013.
DOI: 10.4043/23987-MS

[93] Tilghman BJ. Improvement in
cutting and engraving stone, metal,
glass, etc. US States Patent 108,408.
18 October 1870

[94] Saber KE, Mahmud WM,
Hassan MS. Calculation of EUR form
oil and water production data. In: Paper
Presented at the 8th International
Conference in Industrial Engineering
and Operations Management, Bandung,
Indonesia, March 6-8. 2018

Chapter 3

Hydraulic Fracturing in Porous and Fractured Rocks

Duvvuri Satya Subrahmanyam

Abstract

There are various methods to determine in situ stress parameters, each having its own advantages and limitations. Among the methods available, the hydraulic fracturing method is the most adopted method for in situ stress measurements because of its simplicity and reliability. But the legitimacy of this method becomes questionable in fractured and porous rocks as the amount of experimental work has thus far been limited, especially in the case of its validity in fractured and porous rocks. The relatively slow rates of pressurisation have ensured that when fracture initiation occurs, the sudden increase in volume may lead to a marked drop in pressure in the fractured section, which is easily recognised from the pressure record. This is because pressure cannot be developed if the rate of leakage in the formation is equal to or higher than the flow rate applied for fracture initiation.

Keywords: hydraulic fracturing, fractured rocks, porous rocks, high flow rate, overcoring

1. Introduction

Hydraulic fracturing provides only plane principal stresses, and no information on the other components of the tri-axial stress field is available [1]. In hydraulic fracturing, continuous water pressure is applied in confined area which tends the rock to be tensile and while pressure exceeds the strength of rock, water escapes in weak plane formed [1–15].

Haimson studied on various rock specimens of variable pore pressures. Around 400 specimens have been tested under rational loading conditions. All the specimens failed under tensile manner. He was the one who pointed out the role of water pressure in fracture propagation. His study proved that water pressure increases the pore pressure in turn unable to obtain actual results. The reliability and validity of this method is also questionable when dealing with porous and fractured rocks encountered in underground mines [2–8, 16].

The main objective is to develop a proper and add-on technique for hydraulic fracturing for stress measurement in porous and fractured rocks. Hydraulic fracturing tests were conducted by using different flow rates of water inside the fractured rocks and high viscous fluid in porous strata. The stresses evaluated by this method was correlated with normal flow rate hydraulic fracturing method at the same locations where the rock mass was not fractured, and to circumvent the effect of the porousness, by overcoring technique since porosity does not have any influence on overcoring procedures. The correction factor was introduced during stress evaluation by hydraulic fracturing method in fractured and porous rocks. Normal flow rate is flow rate of fluid during hydraulic fracturing test ranging from 6 to 8 l/min.

This new technique will be helpful in conducting the stress measurements in porous and fractured rocks, which will be highly beneficial to both mining and hydropower related excavations.

The measurement of the state of in-situ rock stress provides essential data for the rational design of underground excavations based on the principles of rock mechanics [13].

The hydraulic fracturing test to determine the stress tensor is rather simple and robust, and it also gives the required magnitude and orientation of the maximum principal stress [17]. Several techniques and equipment have been developed, and are still being developed, to measure this parameter [8].

The main disadvantage of this hydraulic fracturing method when compared to other methods such as overcoring, flat jack and stress-meter, is its limitation when applied to porous and fractured rocks [14, 18]. Rock mass may contain natural occurring discontinuities, including fractures which dissipate fracturing liquid. Hence, it is more difficult to use the hydraulic fracturing process to determine stress conditions in porous and fractured rocks. Whereas in 'non-fractured rock mass', i.e., rock mass without fractures, this limitation is not there. As rocks in a large number of underground coal mines belong to this category, i.e., porous and fractured rocks, finding a methodology to accommodate such rock conditions is essential.

If a high flow rate of fluid is used, experience has shown that there is a tendency of induced fractures to rotate and change the direction of the initial fracture. As the direction of the induced fracture is one of the input parameters for the evaluation of hydraulic fracturing stress, any change in the direction of fracture due to the influence of some external factor, like the flow rate, will give rise to an anomalous pressure or stress value [15].

If, instead of water, a higher viscosity fluid is used for fracture initiation, pressure can be readily developed inside the induced or pre-existing fractures which can be taken for evaluation of stress, but the influence of viscosity on the evaluation of magnitude and direction of stress is not validated.

The above discussed two points show that the limitation in adopting hydraulic fracturing method in porous and fractured rocks is rather due to non-availability of proper technique than the principle of hydraulic fracturing.

2. Hydraulic testing of pre-existing fractures method

Hydraulic testing of pre-existing fractures (HTPF) method provides an evaluation of the complete stress tensor (six components), independent of borehole orientation and material properties.

A portion of a borehole is closed off by use of two inflatable rubber packers adequately pressurized so that they hold on to the borehole wall (**Figure 1**). The water is pumped under continuous flow rate into the portion, gradually increasing the pressure on the borehole wall until a fracture is begun in the rock, or a pre-existing fracture is opened. Pumping is halted, allowing the interval pressure to deteriorate. Several minutes into the shut-off phase, the pressure is released and allowed to return to ambient circumstances. The pressure cycle is repetitive several times maintaining the same flow rate. Key pressure values used in the computation of the in-situ stresses are plucked from the pressure-time record. The repeated cycles deliver redundant interpretations of the key pressures. The attitude of the induced hydraulic fracture, or of the pre-existing fracture, is achieved using an

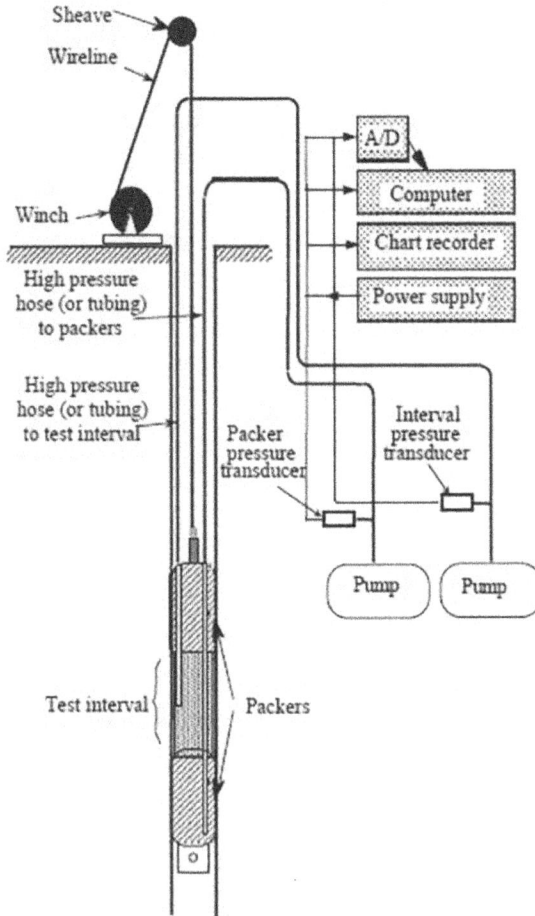

Figure 1.
Typical HTPF test equipment setup.

oriented impression packer. Hydraulic fracturing orientation is related to the directions of the principal stresses [1–15].

HTPF, tests yield an evaluation of the normal stress supported by fracture planes with different known orientations, and the complete stress evaluation results from an inversion of these results.

The main difference between HTPF and Hydraulic fracturing tests are certain assumptions made; otherwise, the process remains same. The following are certain assumptions:

1. There is no theoretical limit to the depth of measurement, provided a stable borehole can access the zone of interest [3].

2. The method assumes that isolated pre-existing fractures, or weakness planes, are present in the rock mass, that they are not all aligned within a narrow range of directions and inclinations, and that they can be instantly opened by hydraulic tests. When the straddled interval includes multiple fractures, it is necessary to verify that only one single fracture has been opened, for the opening of pre-existing fractures change the local stress field [6].

3. Fractures used in stress computations are delineated on the borehole wall under the assumption that their orientation persists away from the hole [5].

4. For a complete stress tensor determination, the method requires a theoretical minimum of six tests.

5. The procedure is applicable for all borehole orientations. It is independent of pore pressure impacts and does not involve any material property determinations.

6. It presumes that the rock mass is consistent within the volume of interest. When tested fractures are isolated from one another by more than 50 m, a hypothesis on stress gradients is essential.

Following are the assumptions in hydraulic fracturing technique:

1. There is no theoretical limit to the depth of measurement, provided a stable borehole can access the zone of interest and the rock is elastic and brittle.

2. Principal stress directions are obtained from the fracture demarcation on the borehole wall under the notion that fracture attitude persists away from the hole.

3. Evaluation of the maximum principal stresses assumes that the rock mass is linearly elastic, homogeneous, and isotropic. It involves considerations of pore pressure effects.

2.1 Hydraulic fracturing in fractured rocks

In Haimson's thesis [10], about 400 tests on hollow cylindrical and cubical specimens of 5 different fractured and non-fractured rocks were conducted under constant tri-axial external loading and increasing borehole fluid pressure [12]. In all of the samples tested, the induced hydraulic fractures were always found to be tensile and no shear failure was observed. The fractures in all the rock types were either vertical or horizontal depending on the applied stresses. These fractures are observed in pairs, mostly parallel to the nearly vertical wellbore axes, and on diametrically opposite sides of the borehole walls [15].

Haimson and Fairhurst [19] showed that the pumped flow increases the pore fluid pressure in fractured/porous formations and produces additional stresses and displacements (**Figure 2**).

Hence it is difficult to get the breakdown pressure (Pc) or the peak pressure in the first pressure cycle in normal flow rates in fractured rocks. Before reaching its peak, pressure typically declines even if pumping is continued at the initial flow rate as the pressure required to induce a hydraulic fracture in HF tests, or fracture reopening in hydraulic fracturing tests on preexisting fractures (HTPF) tests is not sufficient enough. It clearly indicates the following:

1. Critical pressure cannot be reached

2. The slope declines in the pressure–time curve,

3. Declining slope with constant flow rate in subsequent cycles

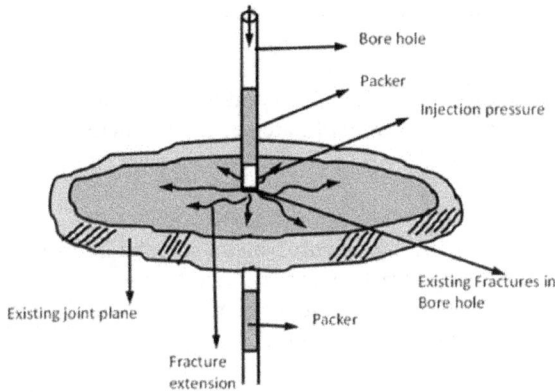

Figure 2.
Existing fractures in a borehole.

4. Shut-in pressure cannot be reached, signifies that maximum fluid is infiltrated in the fractures. (The shut-in pressure denotes at which a hydrofracture pauses generating and closes following pump shut-off. The determination of the shut-in pressure P_{si} is when a sharp break is detected in the pressure-time curve after the initial fast pressure drop following pump shut-off) [9].

In normal conditions or in good rock mass, the shut-in pressure (Psi) reaches, after the pump is shut off following breakdown or fracture reopening. But in the present case in fractured rocks, shut-in pressure cannot be achieved even after repeated cycles. The first difficulty is the pressure decay just before shutting off, and the other difficulty will be not getting shut in pressure to calculate the minimum principal stress [17, 20].

2.2 Difficulties for stress measurements in fractured rocks by various methods

a. The overcoring test method does not permit the testing of rock mass with preexisting fractures with in the test section. The presence of fractures at or near the strain gauges results in erroneous measurements. In addition, the presence of fractures prevents a suitable length section of core being obtained for biaxial testing and determination of the elastic properties of the rock.

b. Flat jack measurements give only induced stress of the area, hence this method is also not suitable.

c. Classical method of hydraulic fracturing test is not suitable as the new fracture cannot be created in the zone of already existing fractures.

d. HTPF method of normal flow rate of water is also not suitable to reopen the existing fractures as the pressure is not sufficient to create the fracture or reopen the existing fracture.

2.3 Solution through innovation

• High flow rate technique with HTPF method can be suitably used in fractured rocks

- High viscous liquid instead of water can also be used for stress measurements in fractured rocks

The above solutions are originally proposed by the author. It will be patented soon at Office of Controller General of Patents, Government of India.

2.3.1 Methodology adopted

1. Hydraulic fracturing measurements were conducted by using different flow rates of water inside the fractured rocks. The stresses evaluated by this method was correlated with normal hydraulic fracturing method at the same locations where the rock mass was not fractured.

2. Hydraulic fracturing measurements were conducted inside the boreholes by using a viscous fluid. The stress evaluation was made using latest software. The stresses evaluated by hydraulic fracturing with viscous liquid method were correlated with stress measured by overcoring method. The stress measured by overcoring method was used as bench-mark for validation as this method does not influence the presence of porous feature in the rock. The correction factor was introduced in the stress evaluation by hydraulic fracturing method in porous rocks.

2.4 High flow rate technique in fractured rocks

In the literature on hydraulic fracture experiments it is generally assumed that a crack will initiate when the tensile stress at the borehole wall exceeds the tensile strength of the rock. It is possible, however, that in regions under tectonic shear stress, shear failure could be induced in the rock about the borehole at much lower fluid pressures than would be required to produce tension cracks, simply by lowering the effective pressure (confining pressure minus pore pressure) to the point where the shear strength of the rock is exceeded [14].

Haimson [11] showed that the compressional strength of the rock mass depends on effective pressure and differential stress. He suggested that a sample subjected to a given confining pressure and differential stress could be made to fail in shear or tension simply by controlling the pore pressure [21]. One way of testing this hypothesis would be to vary the pore fluid injection rate. At slow injection rates the water or any other fluid which is having low viscosity would have time to be drawn-out into the fractured zones and lower the effective pressure, whereas at fast injection rates a steep pore pressure gradient would develop near the borehole. If fluid were pressurized fast enough, even though the shear strength of the rock near the borehole would be surpassed, the load on the area would be supported by the neighbouring rock in which the pore pressure was still low. In this way, shear failure of the sample would not occur and instead, a tension crack would form when the tensile strength of the rock near the borehole was exceeded [7].

In settings with extreme overpressure, pore-water pressure approaches the pressure required for natural hydraulic fracturing. Unlike other fractured seals, hydraulic fractures remain open only if pore pressure exceeds fracture pressure [13].

To test this hypothesis, a series of 24 experiments was conducted at different zones inside the EX-size boreholes (core drilled boreholes of 38 mm diameter) where the rock mass is highly fractured. In all these experiments, the differential stresses were ranging from 10 to 200 bars and the fluid injection rates were varying by 4–16 l/min. It was assumed that the failure mechanisms (shear or tension) observed for different injection rates would be controlled by the pore pressure distribution in each test at the time of failure. The results are validated with normal flow rate of HTPF method in good rock mass zones of the same bore holes. Rock mass quality are characterised using a rating system. The rock mass is categorised into different classes (i.e., very good to very poor), incorporating the combined effects of different geological and geotechnical properties. This enables the comparison of rock mass conditions throughout the site and the delineation of regions of the rock mass ranging from 'very good' to 'very poor', thus providing a map of the boundaries of rock mass quality. The details of the investigations, stress evaluation procedure in fractured rocks and the results are given below.

2.5 Balloon phenomena

At slow bloating rate it is very difficult to inflate the punctured balloon as the air will be leaked through the hole, but at the heavy bloating rate it is possible to inflate the punctured balloon even though the leakage exists. Hence the solution is the rate of bloating should be much higher than the rate of the leakage through the puncture (**Figure 3**). The same balloon phenomena are applicable in the case of hydraulic fracturing testing in fractured rocks. At slow injection rates the fluid would have time to diffuse into the fractures and lower the effective pressure, whereas at fast injection rates a steep pore pressure gradient would develop near the borehole, i.e., If fluid were injected fast enough, even though the shear strength of the rock near the borehole would be exceeded, the load on the sample would be supported by the surrounding rock in which the pore pressure was still low. In this way, shear failure of the sample would not occur and instead, a tension crack would form when the tensile strength of the rock near the borehole is exceeded (**Figure 4**). Hydraulic fracturing is initiated when the fluid pressure exceeds the minimum principal compressive stress by the tensile strength of the host rock. Typical in-situ tensile strengths of rocks are in the order of 0.5–6 MPa (Haimson & Rummel [22], Amadei & Stephansson [23], Enever & Chopra [24]) [2, 19]. The propagation is made possible by the linking up of discontinuities in the host rock ahead of the hydraulic fracturing tip. Discontinuities are significant mechanical breaks in the rock, normally with low or negligible tensile strengths.

2.6 Brief about project area

The experiments have been conducted inside the underground tunnels of proposed underground powerhouse and intake drift area of at Teesta Stage-IV Hydroelectric

Figure 3.
Balloon phenomena similar to hydraulic fracturing test in fractured/porous rocks.

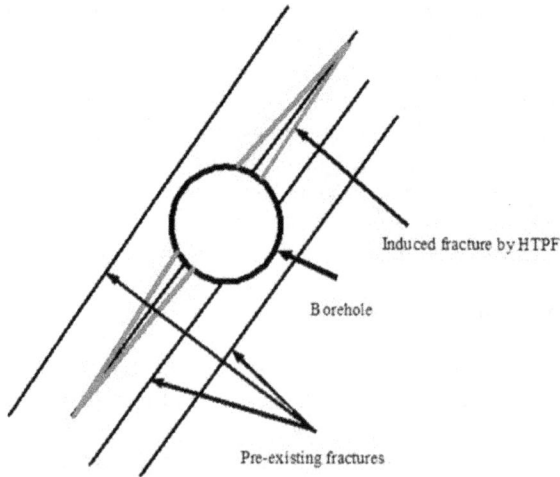

Figure 4.
Induced fracture/reopening of existing fracture in fractured rocks by high flow rate technique.

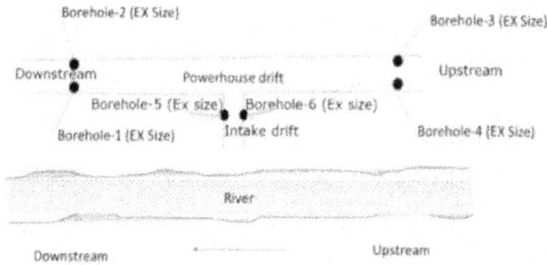

Figure 5.
Configuration of boreholes at powerhouse area-Teesta Stage-IV HEP.

project (**Figure 5**). Teesta-IV Hydroelectric Project was conceptualized in North Sikkim district, Sikkim for harnessing the hydro-power potentiality of Teesta River. The project is located in village Sangklang near Mangan in North Sikkim District.

The geology of the project area is represented by, quartzose phylite with garnet like crystals & ferruginous quartzite. The borehole logging and the cores retrieved from the boreholes are shown in **Figure 6**a–c.

2.7 Investigation procedure

Experiment procedure involves (**Figure 7**) selection of study area in fractured and non-fractured rock mass, conducting hydraulic fracturing tests in study area by

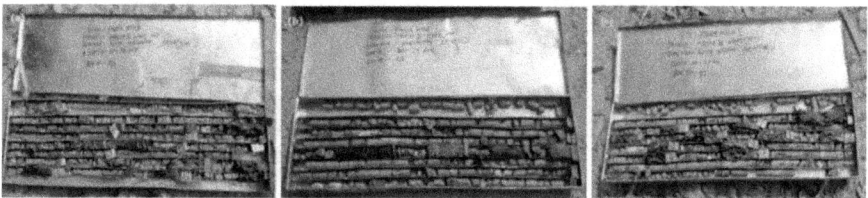

Figure 6.
Cores retrieved from the borehole at (a) and (b) powerhouse area; (c) intake drift.

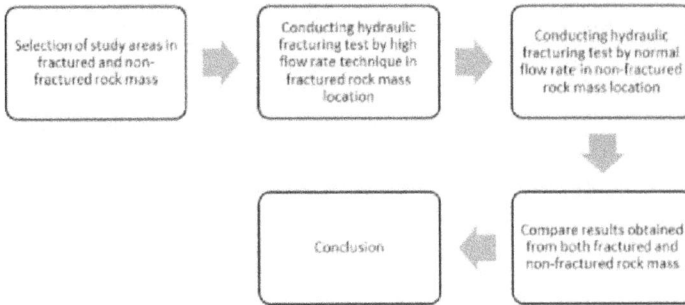

Figure 7.
Flowchart describing methodology and experimental procedures.

high flow rate in fractured rock mass and normal flow rate in non-fractured rock mass. At last, compare the results obtained with fractured and non-fractured rock mass and concluding with results.

After the hydraulic fracturing assembly was positioned at a pre-determined test section where the rock mass is highly fractured (selected based on core inspection **Figure 6**). The back flow from the fracture into the interval section was observed by short valve closures during the venting phase. Finally, the packers were deflated, and tool was moved to the next test section. After all the hydraulic fracturing tests were conducted in all the boreholes, an impression packer tool with a soft rubber skin was run into the holes to obtain information of the orientation of the induced or opened fracture traces at the borehole wall (**Figure 8**).

Experiment 1 (powerhouse upstream)

In trial 1, the experiments were conducted in the EX-size hole at the depth of 10–16 m where the rock mass was completely fractured. This particular zone was selected after careful observation of core logging data. The injection unit was placed at this depth for the pressurization. The pressure was injected at a rate of 6 l/min for a duration of 50–250 sec and the pressure was instantaneously increased up to 50 bars. Critical pressure could not be reached which eventually dropped to zero at the end of cycle. The shut- in pressure could not be achieved even though the pump was shut-off at certain peak levels (**Figure 9**). It clearly indicated that water has

Figure 8.
Tracing of fractures from impression packer at powerhouse downstream area.

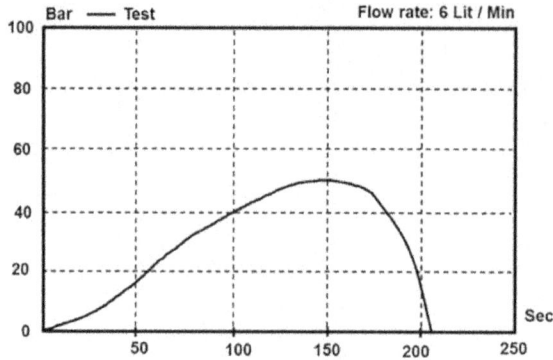

Figure 9.
Experiment 1.

been escaped from the existing fractures and the required pressure could not develop to reopen the fracture. Normal stress required for reopening of the pressure could not build up across the fracture. The pressure time diagram for the flow rate of 6 l/min is given below (**Figure 9**).

Experiment 2

In trial 2, the experiments were conducted in the EX-size hole at the depth of 10–16 m where the rock mass was completely fractured. This particular zone was selected after careful observation of core logging data. The injection unit was placed at this depth for the pressurization of the zone. The pressure was injected at a rate of 9 l/min for a duration of 50–250 sec and the pressure was increased up to 60 bars. Critical pressure could not be reached but there was a decline in the pressure which eventually dropped to zero at the end of cycle. The shut- in pressure could not be achieved even though the pump was shutoff at certain peak levels (**Figure 10**). It clearly indicated that water has been escaped from the existing fractures and the required pressure could not develop to reopen the fracture. Normal stress required for reopening of the pressure could not build up across the fracture. The pressure time diagram for the flow rate of 9 l/min is given below.

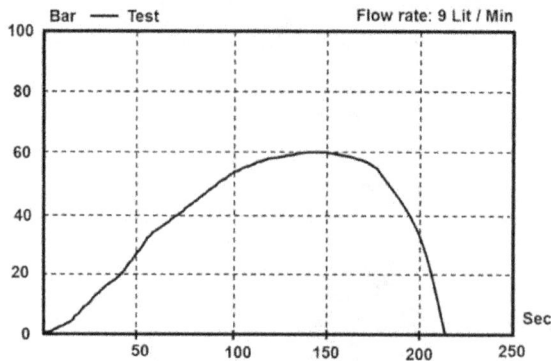

Figure 10.
Experiment 2.

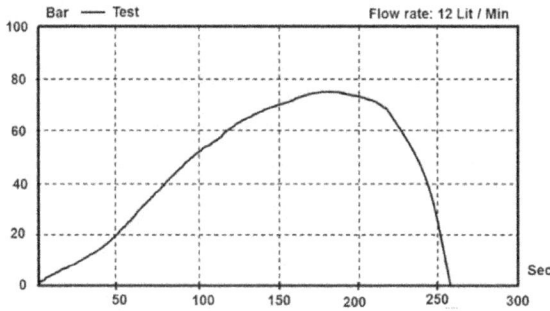

Figure 11.
Experiment 3.

Experiment 3

In trial 3, the experiments were conducted in the EX-size hole at the depth of 10–16 m where the rock mass was completely fractured. This particular zone was selected after careful observation of core logging data. The injection unit was placed at this depth for the pressurization. The pressure was injected at a rate of 12 l/min for a duration of 50–250 sec and the pressure was increased up to 70 bars. Critical pressure could not be reached but there was a decline in the pressure and which eventually dropped to zero at the end of cycle. The shut-in pressure could not be achieved even though the pump was shutoff at certain peak levels (**Figure 11**). It clearly indicated that water has been escaped from the existing fractures and the required pressure could not develop to reopen the fracture. Normal stress required for reopening of the pressure could not build up across the fracture. The pressure time diagram for the flow rate of 12 l/min is given below.

Experiment 4

In trial 4, the experiments were conducted at the same depth of 10–16 m where the earlier experiments were conducted with the flow rate of 6, 9 and 12 l/min. But in this case the flow rate was instantaneously increased to 15 l/min. The pressure was injected at a rate of 15 l/min for a duration of 80 sec and the pressure was automatically increased up to 95 bars. In the first cycle a clear critical pressure could be reached and there was no declining of pressure abruptly. Shut in pressure obtained at 50 bars after the pump was shut off. It clearly indicated that water flow has been required 15 l/min for the existing fractures to reopen.

Where P_{si} is the shut-in pressure, represented as the point of intersection between the tangent to the pressure curve immediately after pump shut-off and that to the late stable section of the pressure curve (**Figure 12**) (Enever and Chopra, 1986). The pressure time diagram for the flow rate of 15 l/min is given below (**Figure 13**).

3. Stress evaluation procedure and results

The in-situ stress measurements were conducted under the following situations:

   i. Influence of topography.

   ii. Presence of anisotropic rock.

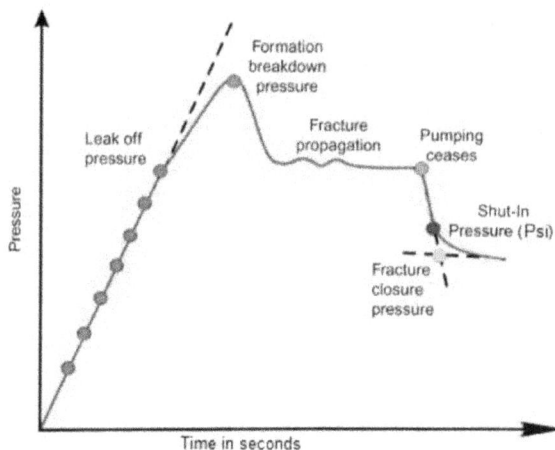

Figure 12.
Shut-in pressure related to Hydraulic fracturing.

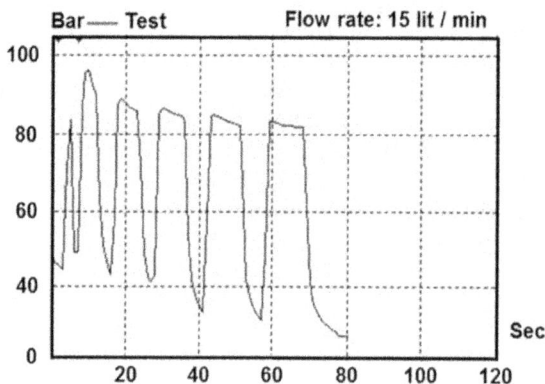

Figure 13.
Experiment 4.

Topography is the study of the land surface and forms the basis for landscapes. For example, topography refers to mountains, valleys, rivers, and craters on earth surface. If a tunnel is being excavated beneath a land consisting of different rock covers or overburden layers, anisotropic conditions are imminent [18].

3.1 Fracture orientation analysis—PLANE

The orientation (strike, dip angle and dip direction) of induced fracture traces obtained from impression packer testing is determined with the program PLANE in consideration of the borehole diameter and orientation [17]. Also, it differs with fracture traces as shown below.

Case I: Vertical borehole—angle from north to mark (0–360 degrees)

Case Ia: Fracture traces parallel to the borehole axis. Distance from mark (reference line) to fracture trace (**Figure 14**).

Case Ib: Inclined fractures (**Figure 15**).

Case II: Inclined borehole—angle from vertical line to reference mark (0–360 degrees)

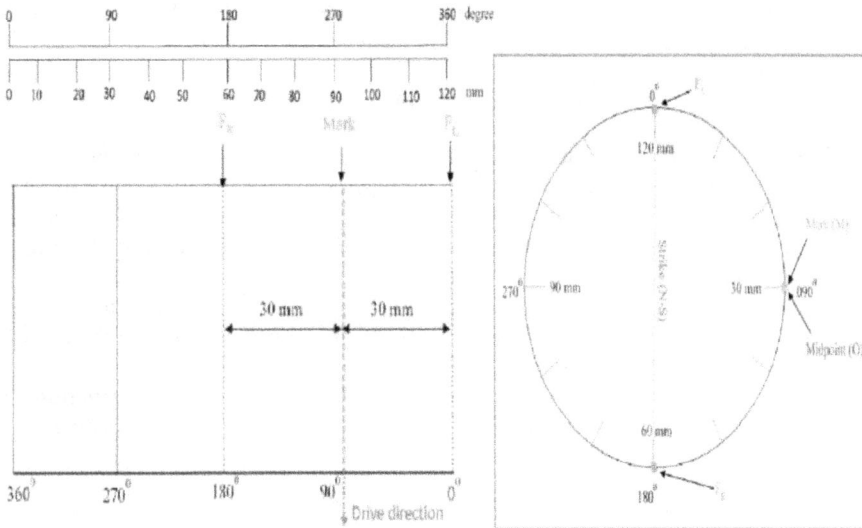

Figure 14.
Trace of the fracture is parallel to the mark found both sides.

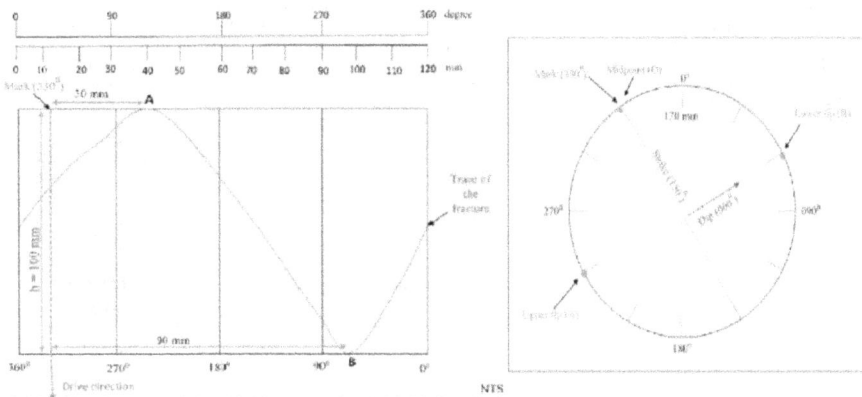

Figure 15.
Inclined fractures.

Case IIa: Fracture traces parallel to the borehole axis.
Case IIb: Inclined fractures

In all cases the result of the calculations is the strike direction (North Over East), the direction of inclination (North over East) and the inclination (90 degrees = vertical) of the fracture plane.

3.2 Data Interpretation code GENSIM

Impression packer tests suggest that in many cases hydraulic testing had been carried out along differently orientated fractures in the rock mass. The interpretation of these data requires sophisticated methods like the GENSIM rather than simple classical hydrofrac hypothesis.

The GENSIM algorithm assumes that the vertical is a principle stress axis and the vertical stress is equal to the weight of overburden rock stress. Stress–depth dependence is neglected within the program through which GENSIM is limited to short depth intervals

$$\sigma_h = (P_{si} - n2.\sigma_v)/(m2 + l2.\sigma_H/\sigma_h) \tag{1}$$

Where, l, m, n in the equation are the cosines of the direction of the induced fracture planes obtained during study; P_{si} is the shut-in pressure in MPa obtained from pressure record for certain depths and σ_H/σ_h is the stress ratio condition which prevails in ground conditions as 1.5, 2 and 2.5 considered for determining the minor principal horizontal stress.

Results in **Table 1** were done by using shut-in pressure data as given in **Tables 2–4** derived from the measurements in the borehole and varying the ratio $\sigma H/\sigma h$ and the strike direction of σH in the horizontal plane.

The stress gradient is plotted to observe any induced stress due to excavation or topography of rock cover (**Figure 16**). It is clearly understood that results determined are free from any influence.

Principal stresses	Intake drift	powerhouse drift (upstream)	powerhouse drift (downstream)
Vertical stress (σ_v) in MPa (calculated with a rock cover 160 m and density of rock = 2.7 g/cc)	4.24	5.08	6.35
Maximum horizontal principal stress (σ_H) in MPa	6.81 ± 1.26	5.46 ± 1.905	8.88 ± 0.855
Minimum horizontal principal stress (σ_h) in MPa	4.54 ± 0.84	3.64 ± 1.27	5.92 ± 0.57
Maximum horizontal principal stress direction	N 20 degrees	N 120 degrees	N 110 degrees
$K = \sigma_H/\sigma_v$	1.60	1.35	1.39

Table 1.
Stress tensors as evaluated at various locations.

Sl. no.	Fracture inclination/dip (degrees) [90 degrees = vertical]	Fracture strike (degrees) [N over E]	Shut-in pressure, P_{si} (MPa)
1	8.1	40	3.7
2	34.4	40	6
3	65.8	22.3	4
4	68.4	159.0	6.3

Table 2.
Fracture orientation data obtained from BH-1 and BH-2 with high flow rate-15 l/min (location: Powerhouse upstream; Teesta stage-IV Hydroelectric project West Bengal).

Sl. no.	Fracture inclination/dip (degrees) [90 degrees = vertical]	Fracture strike (degrees) [N over E]	Shut-in pressure, P_{si} (MPa)
1	72	175	4
2	39	139	5
3	23	26	5
4	64	54	4
5	60	57	6

Table 3.
Fracture orientation data obtained from BH-5 and BH-6 with high flow rate-15 l/min (location: intake drift; Teesta stage-IV Hydroelectric project West Bengal).

Sl. no.	Fracture inclination/dip (degrees) [90 degrees = vertical]	Fracture strike (degrees) [N over E]	Shut-in pressure, P_{si} (MPa)
1	44.2	171	7.6
2	63.2	136.1	6.5
3	77.4	52.6	7.8
4	58.5	156.8	6.0

Table 4.
Fracture orientation data obtained from BH-3 and BH-4 with high flow rate-15 l/min (location: powerhouse downstream; Teesta stage-IV Hydroelectric project West Bengal).

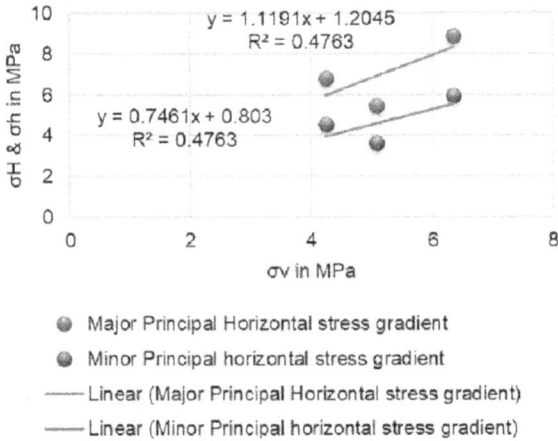

Figure 16.
Stress gradient in fracture rock mass.

4. Hydraulic fracturing test with normal flow rate in same test section of good rock mass

It is necessary to find out or validate the results obtained by using high flow rate technique to measure the maximum principal horizontal stress and its direction in highly fractured rock mass. Hence two to three zones of good rock mass area where the rock mass is not highly fractured were identified in the same borehole and conducted the experiments with normal flow rate method of 4–6 l/min to create the new fracture for classical hypothesis or to reopen the existing fractures. Tests with normal flow rate in non-fractured rock mass would give nearer result to correlate with the high flow rate technique. Other methods may show some difference

All the experiments were conducted in the EX-size hole at the depth where the rock mass was not fractured. These particular zones were selected after careful observation of core logging data. The injection unit was placed at this depth for the pressurization. The pressure was injected at a rate of 6 l/min for a span of 50–250 sec and the pressure was instantaneously increased up to 90–100 bars. The shut-in pressure could be achieved even though the pump was shut-off at certain peak levels [25]. It clearly indicated that normal stress required for reopening of the pressure could build up across the fracture [21]. The detailed procedure of the experiments, results obtained at different places are given below (**Tables 5–8**).

Sl. no	Fracture inclination/dip (degrees) [90 degrees = vertical]	Fracture strike (degrees) [N over E]	Shut-in pressure P_{si} (MPa)
1	8.11	40	3.76
2	34.44	40	6
3	65.88	22.36	4
4	68.46	159.07	6.3

Table 5.
Pressure and fracture orientation data derived from BH-1 and BH-2 with normal flow rate (6 l/min) in the powerhouse drift (upstream); Teesta stage-IV HEP West Bengal.

Sl. no	Fracture inclination/dip (degrees) [90 degrees = vertical]	Fracture strike (degrees) [N over E]	Shut-in pressure P_{si} (MPa)
1	44.28	171	7.7
2	63.27	136.14	6.6
3	77.4	52.65	7.8
4	58.5	156.82	6.0

Table 6.
Pressure and fracture orientation data derived from BH-3 and BH-4 with normal flow rate (6 l/min) in the powerhouse drift (downstream) Teesta stage-IV HEP West Bengal.

Sl. no	Fracture inclination/dip (degrees) [90 degrees = vertical]	Fracture strike (degrees) [N over E]	Shut-in pressure P_{si} (MPa)
1	72.99	175.99	4.2
2	39.06	139.43	5.6
3	23.33	26.46	5.53
4	64.03	54.66	4.6
5	60.8	57.76	6.77

Table 7.
Pressure and fracture orientation data derived from BH-5 and BH-6 with normal flow rate (6 L/min) at Intake drift Teesta stage-IV HEP West Bengal.

Principal stresses	Intake drift	Powerhouse drift (upstream)	Powerhouse drift (downstream)
Vertical stress (σ_v) in MPa (calculated with a rock cover 160 m and density of rock = 2.7 g/cc)	4.24	5.08	6.35
Maximum horizontal principal stress (σ_H) in MPa	7.575 ± 1.47	7.28 ± 2.23	8.95 ± 0.931
Minimum horizontal principal stress (σ_h) in MPa	5.05 ± 0.9803	4.42 ± 1.04	5.97 ± 0.621
Maximum horizontal principal stress direction	N 20 degrees	N 120 degrees	N 110 degrees
$K = \sigma_H/\sigma_v$	1.78	1.35	1.40

Table 8.
Stress tensors as evaluated at various locations.

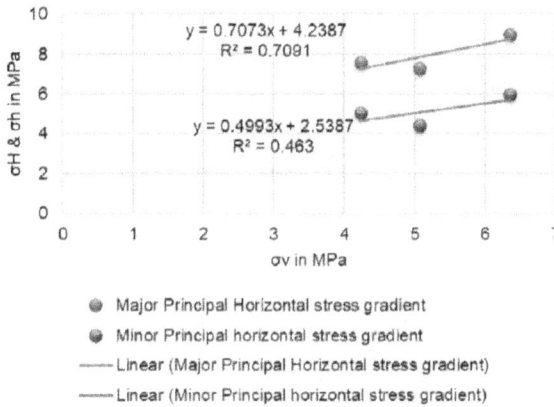

Figure 17.
Stress gradient in non-fractured rock mass.

The stress gradient is plotted to observe any induced stress due to excavation or topography of rock cover (**Figure 17**). It is clearly understood that results determined are free from any influence.

5. Comparison of results obtained from both methods

The hydraulic fracturing tests were conducted by using high flow rate technique in fractured rock mass and normal flow rate technique in good rock mass zones in the same boreholes. A total of 24 hydraulic fracturing tests were attempted in different EX size boreholes inside the tunnels of proposed powerhouse and Intake drift areas. The testing zones selected at the depths between 7 and 27 m. In most hydraulic fracturing testing, at the depth of 7–30 m, pumping rates of 4–6 l/min are sufficient to conduct the entire test. Such pumping rates were sufficient to conduct good hydraulic fracturing tests, but proved to be insufficient for tests in the fractured zones. As this problem became apparent during testing, a high-pressure pump was used in order to achieve higher pumping rates (up to 18 l/min). HTPF method was used for data interpretation and the analysis of the results was done by using PLANE software and GENSIM.

The software PLANE incorporates the impression data with the compass data as input parameters and gives the strike, dip and dip direction as the output known as fracture orientation data.

The software GENSIM computes the stress field based on measured shut in pressure and fracture orientation data. Assumption is that the vertical stress is a principal stress and is equal to the weight of the overburden. The powerful GENSIM program requires only the shut-in pressure and the orientation of an induced or pre-existing fracture. As a result, the role of breakdown pressure and fracture reopening pressure are nil as far as stress computation is concerned [17].

After obtaining the results by both methods it is observed that the direction of maximum principal horizontal stress is not changed. The magnitude of maximum and minimum principal horizontal stresses is also almost same with negligible or fraction of difference. The stress gradients are observed in fractured and non-fractured rock mass. No influence found of any induced stress at any location. The results are compared in **Table 9**.

Stresses	Fractured rock mass	Non-fractured rock mass	Remarks
Maximum horizontal principal stress (σH) orientation	N 20 to N 120 degrees	N 20 to N 120 degrees	No change in orientation
Stress gradient (σ_H/σ_v)	1.19.Z + 1.2 $R^2 = 0.47$	0.7.Z − 4.23 $R^2 = 0.7$	No change in stress gradient
Stress gradient (σ_h/σ_v)	0.74.Z + 0.8 $R^2 = 0.47$	0.49.Z − 2.53 $R^2 = 0.46$	No change in stress gradient

Table 9.
Comparison of results determined in fractured and non-fractured rock mass.

6. Hydraulic fracturing in porous rocks

Hydraulic fracturing method is the accepted technique for measurement of in-situ stresses in hydroelectric projects and in metalliferous mining projects in India and abroad. But its use in coal mines is limited to a few British and Australian coal mines. This is mainly because of the occurrence of porous rocks in coal mines in India and elsewhere.

Scanty literature is available for this type of studies as this procedure of experiments is still in the budding stage. Hence literature references have not been elaborately quoted since it was not an objective of this work to critically compare aspects of our experience with those of other works. The method used in the present study is described here in its near original form in order to place on record the experience gained.

6.1 High viscous fluid technique in porous rocks

The porosity has major effects on hydraulic fracturing technique which results in fracture deviation away from the actual orientation (**Figures 18 and 19**). High viscous liquid instead of water is used for pressurization during hydraulic fracturing, but the influence by using viscous liquid on the stress is not known (**Figures 20 and 21**). The results (**Tables 10–12**) are validated with overcoring technique that is applicable in porous rocks. Over coring method does not get influenced from the presence of porosity in the rock mass [2, 7, 14, 19].

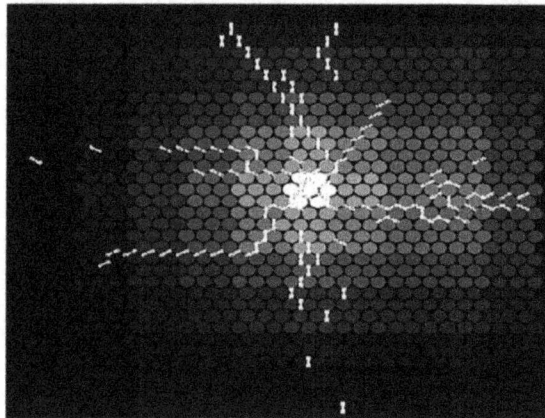

Figure 18.
Hydraulic fracturing test with normal flow rate in porous rocks.

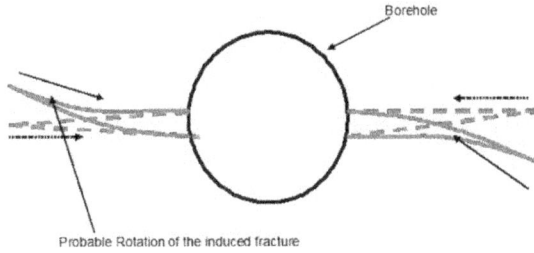

Figure 19.
Hydraulic fracturing test with high flow rate in porous rocks.

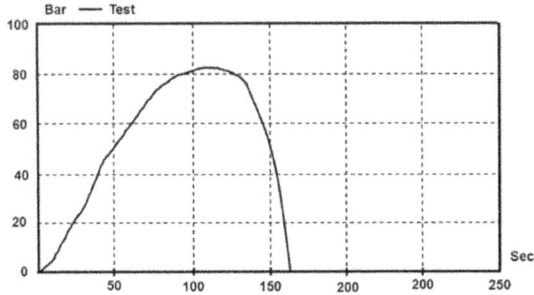

Figure 20.
Pressure drop for low viscous oils (less than 200 cP).

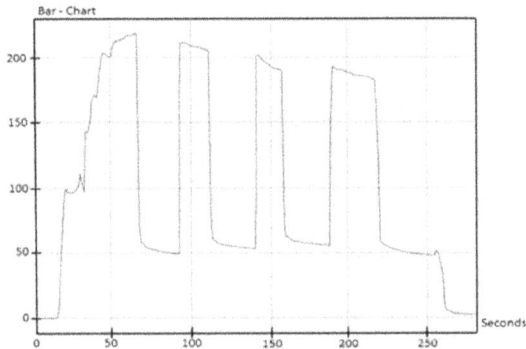

Figure 21.
Pressure-time record obtained using high viscous liquid (ISO VG 320 oil).

Sl. no.	Trace dip (degree) [0–90 degrees]	Trace orientation (degree) [North–East]	Shut-in pressure (Mpi)
1	87	165	9
2	88	160	9.3
3	58	73	8
4	84	128	10.34
5	62	25	5
6	59	84	6

Table 10.
Fracture orientation data obtained from borehole at KTK 8 incline, Singareni Collieries Company Ltd. Telangana.

Sl. No.	Fracture inclination (degrees) [90 degrees = vertical]	Fracture strike (degrees) [N over E]	P_{si} (Mpa)
1	40	25	4.5
2	50	71	8
3	60	20	4.25
4	70	05	5.015

Table 11.
Fracture orientation data obtained from Borehole at Shantikhani longwall mine, Singareni Collieries Company Ltd. Telangana (using high viscous liquid).

Principal stresses	KTK 8	Shantikhani
Vertical stress (σ_v) in Mpa (calculated with a rock cover 160 m and density of rock = 2.7 g/cc)	6.59	11.02
Maximum horizontal principal stress (σ_H) in Mpa	7.31 ± 1.91	11.25 ± 0.4815
Minimum horizontal principal stress (σ_h) in Mpa	3.65 ± 0.957	4.50 ± 0.1926
Maximum horizontal principal stress direction	N 30 degrees	N 20 degrees
$K = \sigma_H / \sigma_v$	1.11	1.02

Table 12.
Stress tensors as evaluated at various locations.

7. In-situ stress measurements by overcoring technique in porous rock mass

Overcoring measurements are common in civil and mining engineering and conducted for design of underground openings [2]. The quality of the measurement depends on quality of drilling, gluing and overcoring, and on the rock characteristics such as anisotropy, discontinuities, and heterogeneity [2, 7, 14, 16, 18, 25].

7.1 Overcoring test procedure

Drilling HX size hole— HX size (150 mm diameter) hole was drilled up to a depth of 7 m in the roof sandstone (**Figure 22**).

Core retrieval—The overcored rock is recovered from the hole using core-breaking chisel that is attached to the rods used for wedging the core off the face (**Figure 23**). An intact length of core (>500 mm) free of fractures and joints is ideal for a satisfactory overcore. The recovered core was considered satisfactory, and free of fractures and voids [16].

Drilling EX size borehole (Pilot hole)—After removal of the HX size core from the borehole, the EX-size Pilot hole (38 mm) was drilled up to 50 cm exactly at the centre of the HX size bore hole from 7 to 7.5 m (**Figure 24**). This hole was collared concentrically with the large diameter hole. To achieve this, the EX-starter barrel is screwed into a stabilizer and about 60 cm of hole drilled. The drill string is then withdrawn and the Pilot hole is drilled with EX twin tube barrel attached to a stabilizer, to a depth of about 60 cm.

When the EX-hole reached the target depth, water was circulated for an additional 10 min so that the drilling sludge and cuttings could be flushed out. The barrel and drill string are then removed and the EX core is recovered for inspection.

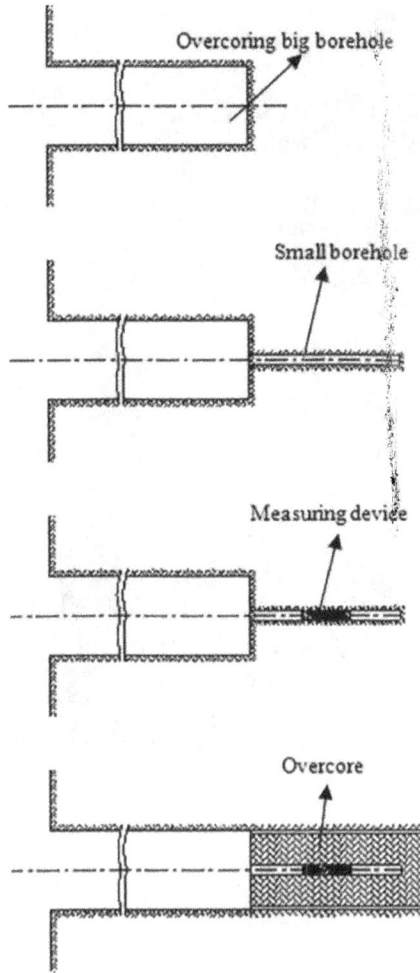

Figure 22.
Overcoming method.

Preparation of glue—It was ensured that the resin and hardener had the correct temperature specification for the expected temperature range. These two were mixed (**Figure 25**) according to the prescribed procedure. Any air pockets remaining were removed by carefully dispersing the glue with a small rod.

Selection of gauge position—The recovered E core was closely inspected to locate the best possible position for the strain gauges. The distance from the strain gauges to any likely weakness planes must be maximized. The other requirement was that the gauges should be at least one diameter away, preferably more, from the other ends of the EX-hole.

Figure 26 show a suitable location of the gauges with respect to the core length. Positioning the gauges too far from the collar can cause problems as the core may break during overcoring and damage the shank and the HI Cell as it tends to rotate in the barrel.

EX size hole measurement—The range of depths that the HI Cell can be placed is limited by the requirement to be approximately beyond two over-core diameters from end of the overcore hole and before the E hole end, minus the stub left when the core is broken out of the hole. Typically, the stub length is up to 150 mm.

Figure 23.
Retrieval of HX size core.

Figure 24.
Drilling of EX size borehole (Pilot hole).

The strain gauge position was measured and it was decided to keep it at 60 cm from the collar of the Pilot hole. The depth to the position where the strain gauges are to be glued, was recorded. The installation rods were marked with tape, which indicates the depth to the end of EX size hole (Pilot hole). The tape was made to coincide with the edge of the collar of the over-core hole.

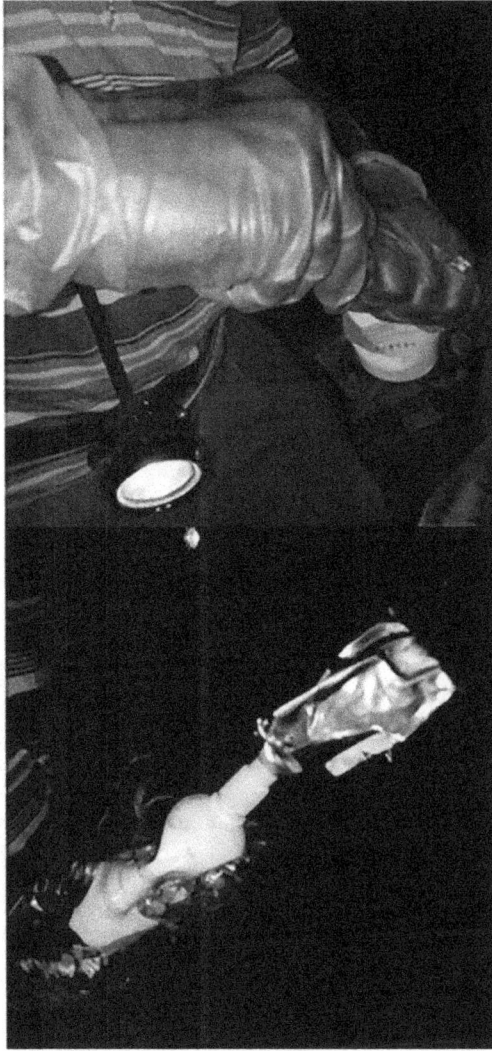

Figure 25.
Preparation of glue.

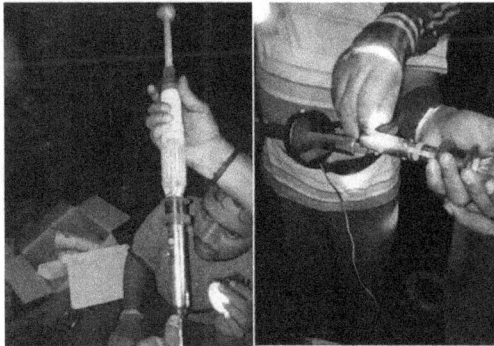

Figure 26.
Selection of gauge position and fixing the pin.

Having determined the distance of the gauges from the EX-size hole (Pilot hole) end, the piston was mounted in the shell at the glue extruded position. The piston rod was then cut to length and taped onto the end of the piston.

Piston attachment—The piston was sprayed with a silicon-based releasing agent to prevent it from bonding to the inside of the gauge shell. The piston is aligned in the Cell as indicated by the scribed lines on the piston and upper rim of the Cell (**Figure 27**). Each of the four holes was lined up, and lead shear pins were placed through the cell wall into the piston.

Installation of HI cell—The completed Cell assembly was inserted in the orienting tool, attached to the trolley (**Figure 28**). The installation tool containing the HI Cell was screwed into the installation rods and the whole assembly pushed up the hole.

Each rod and coupling connections were firmly tightened and the cables were also kept taut. The rods were pushed up the hole until the first tape mark was reached. This indicated that the tip of the piston rod was about to enter the E hole. The Cell was then pushed slowly into the EX-size hole (Pilot hole).

When the second tape mark was reached, the piston rod tip was resting against the end of the Pilot hole. Some extra force was required to break the shear pins and then the rods were pushed slowly inwards so that the glue could evenly distribute itself between the rock and the gauge surface.

In this way, the installation was completed. Once the epoxy glue had gelled and curing was reasonably advanced, the installation rods and the trolley were recovered. Overcoring was commenced after 24 h of installation.

Overcoring—The Cell cable was passed through the centre of each rod, and the rod string was held with a slight tension to ensure that it was not cut by the overcoring barrel (**Figure 29**).

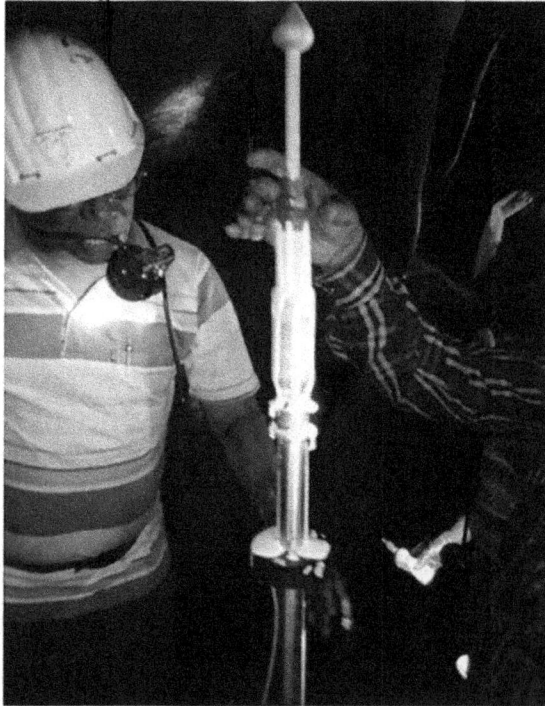

Figure 27.
Piston attachment for CISRO HI cell.

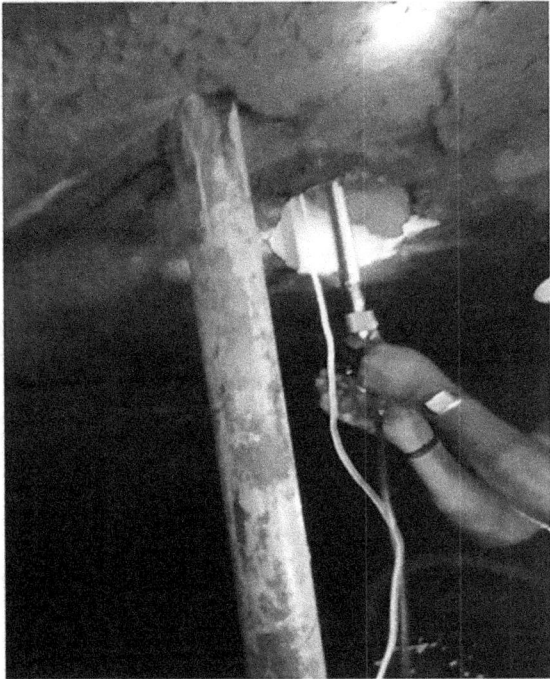

Figure 28.
Installation of CISRO HI cell.

Figure 29.
Overcoring test at KTK 8 Incline.

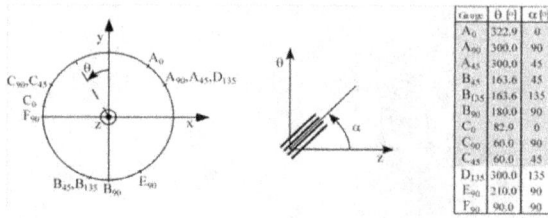

Figure 30.
CISRO HI cell-strain gauge configuration.

The cell contains three strain rosettes 120 degrees apart. The gauge configuration is as follows: two axial, three tangential and four gauges inclined at ±45 degrees in the 9-gauge cell. The 12-gauge cell has one additional 45 degrees and two additional tangential gauges (**Figure 30**). The gauges are 10 mm long and are located 0.5 mm below the outer surface of the cell.

7.2 Determination of strains

Generally, when determining the observed strains from overcoring, a stable value is preferential before overcoring starts and after overcoring stops. The difference among these values is understood to correspond to the strain relief involved in the overcoring process [12]. Generally, each HI-Cell plot shows a peak strain followed by a flat portion that decreases toward the end of the over-core (**Table 13**). The final strains for the site overcore were obtained by averaging the readings in the flat portion of the curve. The recommended overcoring speed of the

Distance (cm)	A0	A90	A45	B45	B135	B90	C0	C90	D135	D135	E90	F90
0	0	0	0	0	0	0	0	0	0	0	0	0
4	0	0	0	0	0	0	0	0	0	0	0	0
10	0	0	0	0	0	0	0	0	0	0	0	0
15	4	3	0	4	6	6	0	0	0	6	0	0
20	6	10	0	10	45	25	35	0	0	55	0	0
25	9	43	−60	35	89	30	55	−50	−15	120	−20	−60
30	18	140	−35	65	95	55	180	−75	−20	220	−45	−90
35	−75	190	50	95	156	220	300	125	−35	280	−100	−114
40	230	220	280	150	320	380	530	580	350	400	390	180
45	235	250	300	432	330	390	540	590	360	410	403	190
50	130	180	150	350	195	250	320	320	250	350	360	150
55	95	90	102	320	190	240	142	220	222	340	220	135
60	98	65	90	280	185	230	95	175	212	250	120	125
65	96	64	95	275	184	220	92	165	150	230	115	124
70	95	63	94	274	183	219	91	163	149	229	114	124
75	94	62	93	273	182	218	90	162	148	228	113	124
80	93	61	92	272	181	217	89	161	147	227	112	124

Table 13.
Change in strain from overcoring HI-Cell.

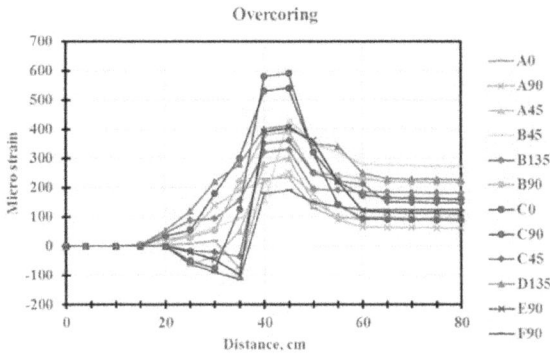

Figure 31.
Overcoring, strains for the KTK 8 incline site.

chuck is 120 rpm to minimize core breakage. **Figure 31** shows overcoring strains as a function of distance for each HI-Cell [14].

7.3 Biaxial testing

In biaxial testing of an overcore section, with no additional instrumentation besides the strain gauges already glued to the inside of the Pilot hole. Measurements can be conducted in the field, in direct unification with overcoring (**Figure 32**). Only radial (biaxial) compression loading is applied to the sample, and there are no restrictions regarding the orientation of the symmetry plane of the rock sample [16]. The biaxial test is exceptionally critical because it establishes the elastic properties of the system, including core, epoxy, and cell, that unlock in situ stress from the over coring strains (**Table 14**). Plots of strain-versus-pressure are shown in **Figure 33**. The calculated elastic properties, namely, Young's modulus and Poisson's ratio are given in **Table 15**.

7.4 Discussion on the results

The calculation of the in-situ state of stress from the measured strains obtained from overcoring measurement in sandstone is based on the theory presented by Amadei [2]. Stress 201X programme allows the user to ascertain stresses and rock properties from raw data output from a CSIRO HI cell and plot overcore (**Figure 31**) and biaxial tests (**Figure 33**). The programme is for genuine ES&S CSIRO HI cells having preset values for alpha and beta angles for the strain gauges. Stress 201X

Figure 32.
Biaxial test.

Pressure, MPa	A0	A90	A45	B45	B135	B90	C0	C90	D135	D135	E90
0	0	0	0	−10	−20	−30	−40	−50	−60	−65	−40
1	10	−10	15	8	−15	−20	−40	−50	−70	−80	−90
3	20	−20	27	15	−17	−26	−50	−80	−100	−110	−120
3.5	25	−30	32	22	−18	−30	−55	−100	−130	−130	−150
4	30	−40	34	28	−19	−40	−60	−110	−130	−140	−160
5.5	40	−50	47	38	−25	−50	−70	−150	−160	−180	−200
7	50	−60	57	48	−35	−60	−90	−200	−220	−240	−260

Table 14.
Change in strain from cyclic biaxial chamber loading to 7 MPa.

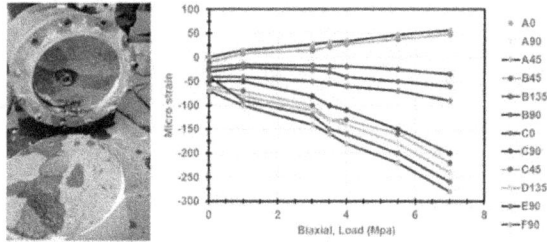

Figure 33.
Plot of micro strain versus pressure.

Sl. No.	Location	Young's modulus, GPa	Poisson's ratio
1	KTK 8	11.6	0.3

Table 15.
Elastic properties calculated from biaxial test.

code was originally developed for CSIR-type triaxial strain cell with a maximum number of 12 strain rosettes with up to 4 strain gauges per rosette. The input values for this program (**Table 16**) are from the averaged circumferential and axial strains for overcore (**Table 13**) and biaxial test results (**Table 15**). The final calculated stress values from Stress 201X program are given in **Table 17**.

Sl.no.	Parameters	Mine data
1	Location	KTK 8–21 incline
2	Hole number	KTK 1
3	Test number	HI 1
4	Hole bearing	—
5	Hole dip	—
6	Date time installed	18-11-2014, 2.00 AM
7	Date time over coring	19-11-2014, 3.00 AM
8	E collar depth	12 cm
9	E hole length	600 m
10	Strain gauge depth	7 m

Sl.no.	Parameters	Mine data
11	Rock temperature	300°C
12	Temperature offset	−0.10°C
13	Drill water temperature	250°C
14	Cell type	CISRO HI Cell
15	Cell number	8069
16	Over core diameter	144.5 mm
17	E hole diameter	38.1 mm
18	Diameter of gauges	35 mm
19	Inner diameter of cell	35 mm
20	Young's modulus of epoxy	2.6 GPa
21	Poison ration of epoxy	0.4
22	K1	1.1258
23	K2	1.2503
24	K3	1.081
25	K4	0.9505
26	Cell gauge factor	2.103
27	Read out gauge factor	2.000
28	Orientation of B90 Gauge (#6)	180
29	Core length	800 mm
30	Maximum biaxial test pressure	15 MPa
31	Rock type	Sandstone
32	Modulus, GPa	11.6
33	Poisons ratio	0.3
34	Maximum temperature change	20°C

Table 16.
Input parameters for software—Stress 201X.

Sl. no	Location	Depth, m	σ_H	σ_h	σ_v	Orientation
1	KTK 8	269	6.8	3.1	6.2	N 30 degrees

Table 17.
Stress values from the ovecoring tests.

7.5 Comparison of the different stress results

The final results of magnitude of the minor and major stress components in the horizontal plane obtained from hydraulic fracturing method show in good agreement with the corresponding stress components obtained from overcoring (**Table 18**).

Two tests were conducted in apparently uniform sandstone by using hydraulic oil as the fracturing fluid. Impression packer images revealed the induced cracks in the reopening pressure. The average horizontal orientation of the cracks obtained at this site shows reasonable agreement with the orientation of the major secondary principal stress component in the horizontal plane obtained from overcoring.

KTK 8–21 incline				
Method of test	σ_H (MPa)	σ_h (MPa)	σ_v (MPa)	Orientation
Hydraulic fracturing	6.59	3.65	7.31	N 30
Over coring	6.8	3.1	6.2	N 30

Table 18.
Comparison of results from the two methods.

8. Discussion and inference

The standard long term instantaneous shut-in pressure revealed reasonable pact with the magnitude of the vertical principal stress component obtained from overcoring at the site. In all the sites tested, this was the only instance in which a viscous fracturing fluid had to be employed specifically to enable a crack to be initiated.

There was no indication of crack spin on the impression packer images.

The results indicate the effect of test fluid viscosity on the ability to reliably estimate the magnitude of minor stress component in the horizontal plane from the long term instantaneous shut-in pressure when crack initiation under a seal is suspected. While the agreement was acceptable, for practical purposes, for the tests conducted with water (especially considering the relatively severe influence of experimental errors at the absolute stress levels involved) the discrepancy in the case of the tests conducted with oil was disproportionate. It was also noticed that the relative differences between the tangent intersection and tangent divergence estimates for instantaneous shut-in pressure decreased as the viscosity of the test fluid intensified.

Re-pressurization of the test zones originally tested with oil or hydraulic oil produced instantaneous shut-in pressures and crack reopening pressures consistent with the results obtained using oil as the only test medium. Testing using a combination of fluids such as this may represent a practical means. These results have important implications in the field wherever the hydraulic fracturing stress measurements are required in fractured and porous rock mass. It is suggested to have a re-look at the long-standing view that the hydraulic fracturing method is not suitable for fractured and porous rock mass. But this study has disproved this assumption. In situ stress may vary from point to point, and method to method in a rock mass, and may have different values when measured over different volumes. Such variations are intrinsic and should not always be seen as anomalies or errors in the measurement themselves and cannot be concluded that no comparison or correlation can be drawn from different methods [18].

9. Conclusions

The work described here however represents the results of field evaluation programme, in which a very pragmatic point of view is being taken. The opportunity is taken to evaluate the results obtained from hydraulic fracturing with the results acquired from overcoring at the same site. The results acquired from overcoring are deemed to deliver a trustworthy indication of the in situ stress field.

The in-situ state of stress is measured for two principal reasons

a. To predict rock response to changed loading conditions caused by construction or excavation, including new engineering procedures that require use of the in-situ stress field as part of the design, and

b. To further understand the tectonic processes.

The hydraulic fracturing stress measurements had become a broadly used technique for determining in situ stresses at depth. It is a technique for understanding rock mass behaviour in conjunction with stability of the excavations in rock. Because of the rapidly expanding use of this method, the method is still evolving in certain rock mass conditions [1, 3–11].

Hence the key objective of this project is to develop a proper methodology for in situ stress measurement by hydraulic fracturing method in porous and fractured rock media, encountered in some of the coal mines as well as in some of the underground tunnels of hydroelectric projects in the Himalayas.

To fulfil the objective of the project, it was proposed to conduct in situ stress measurements in fractured and porous rock mass areas by two different methods at the same location. The hydraulic fracturing stress measurements were conducted by adopting both high flow rate and normal flow rate method in fractured rocks and high viscosity liquid method, and overcoring methods in porous rocks. The stress results by the two methods were correlated with already recognized or established technique as a benchmark. The results of hydraulic fracturing stress measurement methods were authenticated, so that this method can be implemented for stress measurement in porous and fractured rocks and use them widely in mining and hydroelectric projects. This will aid in producing a data bank for in situ stress, which will be highly advantageous for both mining and hydropower industries wherever the rock mass is fractured or porous and the stress measurements are indispensable for designing the support systems.

In the first part of the project, two sites were selected inside a proposed powerhouse tunnel of one of the hydroelectric projects in the Himalayas where the rock formations are fractured. Boreholes were drilled 10–30 m deep depending upon the requirement and site conditions. In situ stresses were measured inside these boreholes by hydraulic fracturing method using manipulation of flow rate. The stress evaluation was made using latest software. The stresses evaluated by this method was correlated with normal hydraulic fracturing method at the same locations where the rock mass was not fractured.

A total of 24 hydraulic fracturing tests were attempted in different EX size boreholes inside the tunnels of the proposed powerhouse and intake drift areas where the rock mass was fractured. The testing zones were selected at depths between 10 and 30 m. In normal conditions, and in good rock mass, the pumping rates of 4–6 l/min are sufficient to conduct the hydraulic fracturing test, but such pumping rates proved to be insufficient for tests in the fractured zones. As this problem became apparent during testing, a high-pressure pump was used to achieve higher pumping rates of up to 18 l/min.

It was observed that with increasing or decreasing pressure in each cycle, the pressure also declined automatically after certain increment of pressure. It is interpreted that, since the flow of water is affected by the whole fractured rock mass, the pressure changes were due to the opening of fractures at different spatial positions.

The hydraulic fracturing tests in good rock mass exposed, repeatable pumping pressures, with the same fracture. This indicates that we were creating a new hydraulic fracturing in a formation which had less tensile strength. Data was evaluated from preexisting reopened fractures, and the orientation of these fractures was analysed to understand how the instantaneous shut-in pressures during the test are related to the value of normal stress across the fracture.

The most reasonable explanation, however, is that at the fast-pumping rate the pressure gradient was so large that the tensile strength of the rock near the borehole exceeded before the shear strength of the outer part of the rock mass was reached.

After shut-off of the pump, instantaneous shut-in pressure was obtained to get the normal stress across the fracture and to calculate the minimum principal stress magnitude and direction.

Stress measurements were conducted by using high viscous liquid in porous rocks; in the same rock mass, at about 1 m away, overcoring method using CSIRO Hollow Inclusion Cell was also carried out. The stresses obtained from hydraulic fracturing method using high viscous liquid were correlated with stresses measured by overcoring method. The stress measured by the overcoring method was used as a benchmark as this method does not suffer from the presence of porosity of the rock.

The average long term instantaneous shut-in pressure showed reasonable agreement with the magnitude of the near vertical principal stress component obtained from overcoring at the site. This was the case in which a viscous fluid had to be employed specifically to enable a crack to be initiated, and the shut-in pressure used to make estimates of some stress component magnitudes.

The results indicated the effect of test fluid viscosity on estimation of the magnitude of minor horizontal stress components. It was observed that the relative differences between the tangent intersection and tangent divergence estimates for instantaneous shut-in pressure decreased as the viscosity of the test fluid increased.

10. Summary

Hydraulic fracturing method is the accepted technique for measurement of in situ stresses in hydroelectric projects and in metalliferous mining projects in India and abroad. But its use in coal mines is limited to a few British and Australian coal mines. This is mainly because of the occurrence of porous rocks in coal mines in India and elsewhere.

Despite the extensive theoretical work on the subject of hydraulic fracturing that had been carried out by the mid-1960s, it is for only restricted for fractured rocks. Extensive studies couldn't provide proper solution for conducting hydraulic stress measurements, where it is difficult to conform on the legitimacy of the results [1–15].

The stress measurements in coal mining areas are determined using overcoring methods. Though porosity of the rock mass does not have influence on the stress measured using this method, due to workable limitations, it can be used for shallow depth only. However, the need for the stress measurements at the deeper levels is essential for proper planning of layouts etc. Therefore, it has been widely accepted that hydraulic fracturing technique will be suitable for porous media also provided the practical limitations are overcome.

To fulfil the objective,

a. The hydraulic stress measurement has been conducted by adopting high flow rate method in fractured rocks.

b. The in-situ stress results have been compared with the in-situ stress results by adopting normal flow rate obtained in the same test section of good rock mass condition.

c. The results of hydraulic fracturing can be validated, and the method can be adopted for in-situ stress measurement in fractured rocks.

d. The hydraulic stress measurement has been conducted by using high viscous liquid in porous rocks.

e. The in-situ stress results have been compared with the in-situ stress results by overcoring method obtained in the same test section of porous rock mass condition.

f. The results of hydraulic fracturing can be validated, and the method can be adopted for in-situ stress measurement in porous rocks.

Author details

Duvvuri Satya Subrahmanyam
National Institute of Rock Mechanics, Bangalore, India

*Address all correspondence to: subbu3268@gmail.com

IntechOpen

References

[1] Baumgartner J, Rummel F. Experience with 'fracture pressurization tests' as a stress measuring technique in a jointed rock mass. International Journal of Rock Mechanics, Mining Sciences & Geomechanics. Abstracts. 1989;**26**:661-671

[2] Amadei B. Applicability of the theory of hollow Inclusions for overcoring stress measurements in rock. Rock Mechanics and Rock Engineering. 1985;**18**:107-130

[3] Cornet FH. Stress determination from hydraulic tests on preexisting fractures—The HTPF method. In: Proc. Int.Symp.on Rock Stress and Rock Stress Measurements. Stockholm, Lulea, Sweden: Centek Publ.; 1986. pp. 301-312

[4] Cornet FH, Burlet D. Stress field determinations in France by hydraulic tests in boreholes. Journal of Geophysical Research. 1992;**97**: 11829-11849

[5] Cornet FH, Julien P. Stress determination from hydraulic test and focal mechanisms of induced seismicity. International Journal of Rock Mechanics, Mining Sciences & Geomechanics Abstract. 1989;**26**: 235-238

[6] Cornet FH, Valette B. *In situ* stress determination from hydraulic test data. Journal of Geophysical Research. 1984; **97**:11527-11537

[7] Enever JR. Ten years' experience with hydraulic fracture stress measurement in Australia. In: Proc. of the Second International Workshop on Hydraulic Fracture Stress Measurements; Minnesota: 1988. pp. 1-92

[8] Evans K. Some problems in estimating horizontal stress magnitudes

in thrust regimes. In: Proceedings of the second International Workshop on Hydraulic Fracturing Stress Measurements. Vol. 1. 1988. pp. 275

[9] Gronseth JM, Key PR. Instantaneous shut in pressure and its relationship to the minimum *in situ* stress. Hydraulic fracturing stress measurements. In: Proceedings of a Workshop December 2-5; 1981

[10] Haimson BC. Hydraulic fracturing in porous and nonporous rock and its potential for determining *in situ* stresses at great depth [unpublished PhD thesis]. University of Minnesota; 1968. pp. 234

[11] Haimson BC. Earthquake related stresses at Rangley. Colorado. In: Proceedings of 14th US symposium of Rock Mechanics. University Park. ASCE; 1973. pp. 689-708

[12] Price Jones A, Whittle RA, Hobbs NH. Measurement of *in situ* rock stresses by overcoring. Tunnels and Tunneling. 1984. p. 12

[13] Rummel F. Fracture Mechanics Approach to Hydraulic Fracturing Stress Measurements. Fracture Mechanics of Rocks. London: Academic Press; 1986. pp. 217-239

[14] Sjoberg J, Christiansson R, Hudson JA. ISRM suggested methods for rock stress estimation—Part 2: overcoring methods. International Journal of Rock mechanics and Mining Sciences. 2003;**40**:999-1010

[15] Worotnicki G, Walton RJ. Triaxial hollow inclusion gauges for determination of rock stresses *in situ*. In: Supplement to Proc. ISRM Symp. on Investigation of Stress in Rock, Advances in Stress Measurement. Suppl.1-8. Sydney, Australia: The Institution of Engineers; 1976

[16] Dean AK. Manufacturing Procedures for the tri-axial Hollow Inclusion Stress Gauge. C.S.I.R.O. Technical Report; 1978

[17] Rummel F, Hansen J. Interpretation of hydrofrac recordings using a simple fracture mechanics simulation model. International Journal of Rock Mechanics, Mining Sciences & Geomechanics Abstract. 1989;**26**: 483-488

[18] Stephansson O. State of the art and future plans about hydraulic fracturing stress measurements in Sweden. In: Proceedings of Hydraulic fracturing stress measurements. Monterey. Washington, DC: National Academy Press; 1983. pp. 260-67

[19] Haimson BC Fairhurst C. *In situ* stress determination at great depth by means of hydraulic fracturing. Proceedings of 11th US symposium of Rock Mechanics. Berkeley: SME/AIME; 1970. pp. 559-584

[20] Haimson BC, Fairhurst C. Initiation and extension of hydraulic fractures in rocks. Society of Petroleum Engineering Journal. 1967;7:310-318. DOI: 10.2118/1710-PA

[21] John D, McLennan JC. Do instantaneous shut-in pressure accurately represent the minimum principal stress. Hydraulic fracturing stress measurements. In: Proceedings of a Workshop; December 2-5; 1981

[22] Haimson BC, Rummel F. Hydrofracturing stress measurements in the Iceland Research Drilling Project drill hole at Reydarfjordur, Iceland. Journal of Geophysical Research. 1982: 87. DOI: 10.1029/JB087iB08p06631. ISSN: 0148-0227

[23] Amadei B, Stepahnsson O. Rock Stress and Its Measurement. Chapman & Hall Publisher. 1997. ISBN 13: 9780412447006.

[24] Enever JR, Chopra PN. Experience with hydraulic fracture stress measurements in granite. Proceedings of the International Symposium on Rock Stress and Rock Stress Measurement. Stockholm, Sweden: Centek Publishers; 1986. pp. 411-420

[25] Stephen H, Hickman M, Zoback D. The interpretation of hydraulic fracturing pressure-time data for *in situ* stress determination. Hydraulic fracturing stress measurements. In: Proceedings of a Workshop December 2-5; 1981

Chapter 4

Hydraulic Fracture Conductivity in Shale Reservoirs

Javed Akbar Khan, Eswaran Padmanabhan and Izhar Ul Haq

Abstract

Optimum conductivity is essential for hydraulic fracturing due to its significant role in maintaining productivity. Hydraulic fracture networks with required fracture conductivities are decisive for the cost-effective production from unconventional shale reservoirs. Fracture conductivity reduces significantly in shale formations due to the high embedment of proppants. In this research, the mechanical properties of shale samples from Sungai Perlis beds, Terengganu, Malaysia, have been used for computational contact analysis of proppant between fracture surfaces. The finite element code in ANSYS is used to simulate the formation/proppant contact-impact behavior in the fracture surface. In the numerical analysis, a material property of proppant and formation characteristics is introduced based on experimental investigation. The influences of formation load and resulted deformation of formation are calculated by total penetration of proppant. It has been found that the formation stresses on both sides of fractured result in high penetration of proppant in the fracture surfaces, although proppant remains un-deformed.

Keywords: shale, hydraulic fracturing, proppant embedment, contact analysis, fracture conductivity

1. Introduction

The purpose of injecting proppants in shale reservoirs is to maintain the fracture conductivity for a longer period and to prevent the fracture from closure due to subsurface stresses. On the other hand, the proppants themselves can be a problem in the case where they develop surface penetration in the formation. As a result, the proppant is embedded into the formation and decreases the fracture conductivity of the reservoir as shown in **Figure 1**. Due to inhomogeneous stress distributions between quartz grains and proppants, high tensile stress concentrations beneath the area of contact between quartz grains and proppants are observed even at small external stress applied to the rock-proppant system. These high-stress concentrations are responsible for the early onset of damage at the fracture face and determine the type of proppant failure [1].

Water imbibition and some other tests on saturated shale were carried out to observe the crack generation process and compare the failure patterns as well as damage resistance of saturated shale kernels and unsaturated shale kernels. The average damage resistance of saturated kernel water is found to be around 11.69 MPa compared to 30.57 MPa of unsaturated shale kernels, which implies that water can decrease the resistance to shale damage and helps in generating

IntechOpen

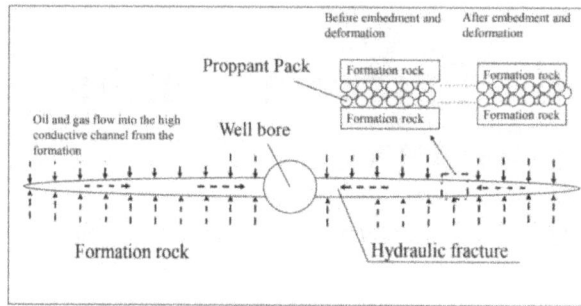

Figure 1.
Propped hydraulic fracture conductivity [2].

fractures [3]. Fracture networks created during the process of hydraulic fracturing usually have a complex pattern. Most of these fractures are kept open by the incorporation of proppants in the form of proppant packs, as shown in **Figure 1**. In the case of secondary fractures, other than bi-wing fractures, proppants are unable to enter into the fractured surface due to narrow apertures and thus, these fractures cannot maintain conductivity for a longer period. The effective vertical and horizontal stresses are responsible for the decrease in hydraulic fracture conductivity and an estimated 60% decrease in propped fracture conductivity occurs by increasing effective stress from 6.2 to 34.48 MPa [4].

Considering the narrow apertures of secondary hydraulic fractures, a partial monolayer of proppant, that is, a single layer of proppant having uneven distribution of proppants over the fracture surface, can be introduced instead of multilayer proppant to maintain the maximum possible conductivity for production improvement [5]. The variation in the aperture and surface roughness of the hydraulic fractures are considered as main reasons for the uneven distribution of proppants. In this regard, the study of the conductivity of fractures with narrow apertures, filled with a monolayer of proppant, can be used for the optimization of hydraulic fracturing and the analysis of production in shale reservoirs. In the past, various types of compression, such as long-term and short-term compression on a single proppant, have been studied in-depth by diametric compression tests and DEM/FEM simulations. Most proppants have shown creep behavior under long-term compression [6]. **Figure 2** shows that the embedment potential is related to many factors especially the proppant material, shape, concentration, and ability of the proppant to resist sinking in the fracture zone [9]. During hydraulic fracturing treatment, high fluid velocities in the fracture are generated by the small contact area between the wellbore and fracture, which results in the erosion of the proppant and fracture connectivity [10].

A computational fluid dynamics study with Eulerian granular modeling (EGM) that is based on solid pressure model and kinetic theory indicates that the transport of the proppant in complex fracture geometries is significantly affected by the dynamics of the fracturing fluids and the properties of the proppant [11]. According to parametric studies, a higher injection rate and lightweight proppants are beneficial for the transport of the proppant through the fracture junctions and to carry proppant in hydraulic fractures and natural fractures [11]. A DEM-CFD (discrete element method–computational fluid dynamics) and the experimental study indicate that during the closure period, the height of the proppants pillar decreases and diameter increases [12]. The proppant flowback could occur easily with a large proppant pillar height or a large fluid pressure gradient. However, the higher bonding strength of the fibers results to improve the stability of the proppant pillar [12].

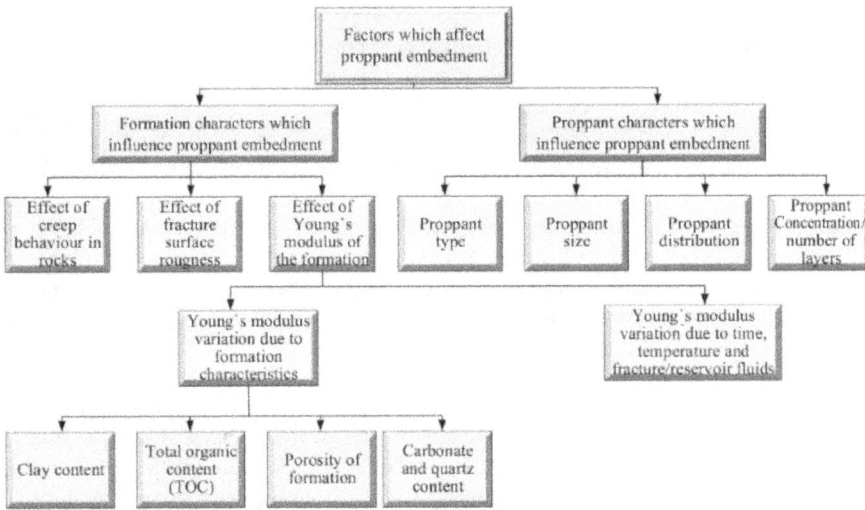

Figure 2.
Flowchart showing factors responsible for embedment [7, 8].

Proppant pillar is defined as concentrations of proppant in the form of pillars that maintain the aperture of the hydraulic fractures. The change in the optimum distance which is defined as the distance between proppant packs that has the potential to maintain the maximum conductivity after proppant embedment under a sparse distribution condition is primarily controlled by closure pressure, the rock's elastic modulus, and the proppant elastic modulus. It also states that the proppant concentrations and the poroelastic effect do not influence this optimum distance [13]. Studies based on analytical and discrete element method (DEM) have led to the understanding of the effects of various factors such as proppant size combination, concentration, time ratio, elastic modulus-to-stress ratio, and looseness coefficient [14, 15]. In these studies, deformation was considered elastic; however, actual phenomena can be captured by considering the intermediate states of elasticity and plasticity such as elastoplastic behavior of rock as well as proppant. In the case of monolayer proppant distribution, the embedment depth and contact stress decrease with the increase in proppant concentration [16]. In the past, machine learning and computational fluid dynamics approaches have been used to explore the well operation and the transport of sand particles by the injection of foam [17–23].

The production performance of fractured wells depends on two factors, that is, formation parameters and fracture parameters [24]. Formation parameters include porosity, permeability, and geo-mechanical properties of the formation, while the fracture parameters comprise a length, aperture, and conductivity of fractures [25, 26]. Hydraulic fracture conductivity reflects the transport capacity of the permeable channel through the reservoir and any alteration to this permeable channel will directly impact the stimulation achieved from the fracturing treatment [27]. The experimental study performed on shale samples with fluids shows that the reduction in the elastic modulus can lead to a significant reduction in the effective fracture conductivity [28]. Zhang et al. reported an 88% reduction in fracture conductivity by injecting water at 27.58 MPa closure stress [29]. In this study, water as a fracturing fluid has been injected to find the excessive proppant embedment caused by the interaction of water with shale matrix, altering the hydraulic fracture conductivity. Water injection increased of local pore pressure and reduction of bonding strength of mineral in clay-rich shale that led to the softening of shale.

The effects of rock stiffness, the roundness of proppant, and the effective stresses on the conductivity of the fracture were studied by a geomechanics-fluid mechanics-coupled numerical workflow considering the interaction between rock matrix and proppant as well as fluid flow in a hydraulic fracture during the process of the reservoir depletion [30]. Compared to the weak shale, less embedment of proppant is observed in sandstone, having high stiffness, which indicates that the rock matrix with higher stiffness is helpful in maintaining the fracture aperture and conductivity [30]. The correlation between the fracture conductivity and the corresponding production performance was quantitatively analyzed using the finite element method [31]. The proposed research can provide valuable information on the unconventional maximization of resource recovery [31]. For future extensions, a network of fractal fractures with a stochastic-based fractal fracture network combined with micro-seismic events can be coupled to quantify the complex fractures of the network to improve fracture conductivity and production performance [31].

The hydraulic fracturing process is a costly job; therefore, improving the reservoir quality evaluation mechanism and optimization of the technical parameters are important. As the low-quality hydrocarbon (HC) shale reservoirs are also gaining increasing attention, to this end, optimization of the hydraulic fracture conductivity is of utmost importance to make the job profitable. To estimate the actual hydraulic fracture conductivity in shale reservoirs, the computational contact analysis of proppant between fracture surfaces has been carried out in this study. In the numerical analysis, a material property of proppant and formation characteristics is introduced from the experimental analysis. The influence of formation load and resulting deformation of formation is calculated by the total penetration of proppant. The deformation mechanism and proppant embedment in shale rocks, saturated with fracking fluid, are then simulated. The finite element code in ANSYS is used to simulate the shale reservoir/proppant contact-impact behavior in the fracture surface. The embedment depth of the shale samples was obtained by numerical as well as experimental methods and the permeability was calculated by the Kozeny–Carman correlation.

2. Procedure

In this section, steps to measure embedment and fracture conductivity of fractured shale have been presented. Governing equations of numerical simulation to study the impact of embedment on the reduction of fracture conductivity have been presented in Section 2.3. The conductivity reduction due to embedment was modeled with a computational fluid dynamics (CFD) approach. We used the CFD software package CFX (ANSYS Inc.) in this work and applied the boundary conditions as shown in **Figure 3** and **Table 1**. Initially, geotechnical characteristics of shale formation were calculated to define these properties in the software.

2.1 Elastic geo-mechanical properties

Dynamic elastic properties of the shale lithofacies, that is, Young's modulus and Poisson's ratio were calculated using compressional and shear velocities measured on shale core samples. The following equations [32] are used to obtain the respective values.

$$\text{Dynamic Young Modulus } (\nu) = \frac{P_b.V_s^2\left(3V_p^2 - 4V_s^2\right)}{V_p^2 - V_s^2} \qquad (1)$$

$$\text{Dynamic Poisson's Ratio (E)} = \frac{V_p^2 - 2V_s^2}{2\left(V_p^2 - V_s^2\right)} \qquad (2)$$

where V_s and V_p represent the shear and dynamic wave velocities in Km/s and P_b is the bulk density of the shale in gm/cc. The results (**Table 2**) show that massive siliceous shale has a high value of Young's modulus and a low Poisson's ratio in

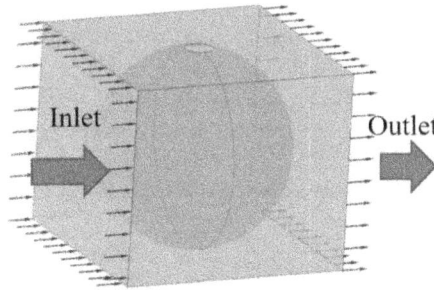

Figure 3.
The direction of flow at inlet and outlet across the proppant.

Injected medium	Water, dynamic viscosity: 0.890 [cP], reference temperature: 25 [C], normal speed: 0.5 [m s^{-1}]
Analysis type	Steady state
Boundary conditions	
Wall	Wall roughness: smooth wall
Inlet	Flow regime: subsonic Turbulence = medium intensity and eddy viscosity ratio, it is the ratio between the turbulent viscosity and the molecular dynamic viscosity. Eddy viscosity ratio is often also called turbulent viscosity ratio or simply viscosity ratio.
Outlet	Flow regime: subsonic Relative pressure = 0 [Pa] Pressure averaging: average over whole outlet
Domain models	Buoyancy model: non-buoyant Morphology: continuous fluid Turbulence model: SST is used which is the shear stress transport turbulence model, it is a widely used and robust two-equation eddy-viscosity turbulence model used in computational fluid dynamics. The model combines the k-omega turbulence model and K-epsilon turbulence model such that the k-omega is used in the inner region of the boundary layer and switches to the k-epsilon in the free shear flow. Convergence control: length scale option = conservative Maximum number of iterations = 100 Subsystem: momentum and Mass, U-Mom, V-Mom, W-Mom, P-Mass, Subsystem: Turb-KE which is turbulence kinetic energy, it is defined as half the sum of the variances (square of standard deviations) of the velocity components and Turb-Freq which is turbulent frequency here infers to the vortex shedding frequency (the frequency with which the vortices are shed behind the proppant during fluid flow).
Assumptions	(1) The proppants are rigid body; (2) monolayer/single proppant is used in the embedment and conductivity analysis. In addition, deformation is calculated in the formation, while no deformation proppant only penetration in the formation is considered; and (3) the entire simulation is isothermal.

Table 1.
Basic input parameters, conditions, and assumptions.

Lithofacies name	Young's modulus (GPa) (dynamic)	Young's modulus (GPa) (static)	Poisson's ratio (dynamic)	Poisson's ratio (static)	Average Young's modulus (GPa)	Average Poisson's ratio
Massive Argillaceous Shale (MAS 1)	30	09	0.33	0.25	19.5	0.29
Massive Argillaceous Shale (MAS 2)	32	12	0.35	0.23	22	0.29
Massive Siliceous Shale (MSS)	46	16	0.27	0.15	31	0.21

Table 2.
Elastic and strength properties of the lithofacies identified in this study.

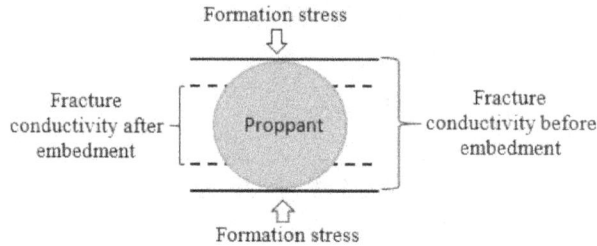

Figure 4.
Reduction in the fracture conductivity by embedment in the fracture surfaces, modified from Zhang et al. [29].

comparison with massive argillaceous shale facie. Static Young's modulus and static Poisson's ratio were measured on cylindrical core samples of 2.0 in diameter and 4.0 in length. The specimens were failed in a triaxial setup and the deformations (axial and radial strains) were measured on the installed strain gauges. The static elastic parameters were calculated using the following equations.

$$\text{Static Young Modulus } (E) = \frac{\sigma_a}{\varepsilon_a} \tag{3}$$

$$\text{Static Poisson's Ratio } (\nu) = -\frac{\varepsilon_r}{\varepsilon_a} \tag{4}$$

where σ_a represents axial stress, while ε_a and ε_r represent axial and radial strains measured on the samples under deformation. Like dynamic parameters, the static Young's modulus of the massive siliceous shale is also greater than the static Young's modulus of argillaceous shale facie. Overall, values obtained for the static moduli are less than the dynamic moduli values measured on the same samples (**Table 2**). This is per previous findings done on producing shale formations. The correlation between dynamic and static moduli is shown in **Figure 4**. Strength parameters of the lithofacies are calculated using the equations below.

$$\text{Shear Modulus } (G) = \frac{E}{2(1+v)} \tag{5}$$

$$\text{Plane Strain Modulus } (E') = \frac{E}{1+v^2} \tag{6}$$

2.2 Calculation of embedment depth

The factors responsible for the change in fracture width and conductivity after hydraulic fracturing in shale reservoirs include proppant embedment and proppant deformation. The embedment of the proppant involves the penetration of the proppant inside the fracture surface, while proppant deformation is directly related to the strength of the proppant [33]. The deformation of the proppant can be described by Eq. (7), as the proppant can be assumed as an elastic body while the penetration of proppant into a body can be solved using contact mechanics [34]. The contact problem can be formulated as a constrained minimization problem, where the objective function to be minimized is the total potential energy (Π) of the bodies in contact. The energy for this system can be written as

$$\Pi(u) = \frac{1}{2}ku^2 - fu \qquad (7)$$

where k is the stiffness matrix, u is the displacement field, and f is the external force. Several constrained minimization algorithms can be used to solve the problem of the equation, such as the penalty method, the Lagrange multipliers method, and the augmented Lagrangian method. The results presented in this paper are based on the augmented Lagrangian method according to the ANSYS implementation. Augmented Lagrangian methods are a certain class of algorithms for solving c onstrained optimization problems. They have similarities to penalty methods in that they replace a constrained optimization problem with a series of unconstrained problems and add a penalty term to the objective; the difference is that the augmented Lagrangian method adds yet another term, designed to mimic a Lagrange multiplier. The augmented Lagrangian is related to, but not identical with the meth od of Lagrange multipliers. A general discussion of these techniques can be found in the literature on contact mechanics [35–37]. In this study, a numerical model is developed based on the experimental investigation carried out on the embedment of 20/40 mesh proppant (size between 20 and 40 μm) on shale samples from Sungai Perlis beds, Terengganu, Malaysia. The core samples were subjected to uni-axial stress of 20 MPa to find the proppant embedment in the formation, as shown in **Figure 4**. Such a high compression force was applied to investigate the embedment under reservoir conditions.

The embedment cell consists of a transparent cylindrical tube where a shale sample is placed. A metal loading ram is used to load the shale-proppant stack and deformation is measured as the axial load is increased. The deformation in shale, assuming elastic behavior, is quantified using Young's modulus and the applied load. **Figure 5** shows the universal testing machine (UTM) used for measuring

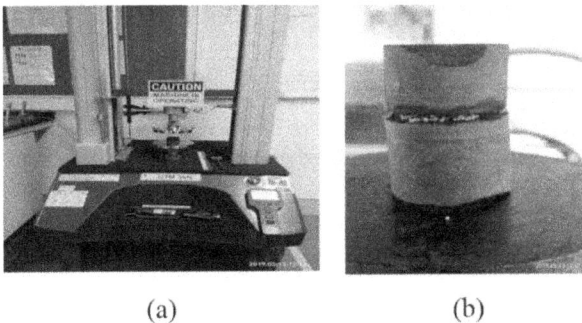

(a)   (b)

Figure 5.
Embedment by compression (a) embedment test by bi-axial compression and (b) sample under UTM machine.

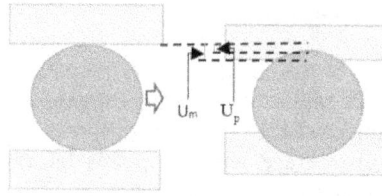

Figure 6.
Illustration of embedment of the surface due to external load.

compression and tensile strengths of materials and samples used in the proppant embedment tests.

In this study, the contact behavior of proppant and rock was carried out using structural mechanics. Experimental boundary conditions are implemented to find the impact of load on proppant penetration in the rock surface. The resulting properties from experiments are introduced in a finite element model to find the effect of fracture surface on the embedment of proppant, as shown in **Figure 6**. A finite element study with ANSYS Workbench has been performed for the computational contact analysis [38]. In a subsurface reservoir, proppants experiences compression from both sides in the formation; therefore, biaxial test should be carried for precise estimation of embedment. Proppants inside the fracture also contact each other due to subsurface stresses. Due to external force, the stress–displacement relationship is as follows:

$$\rho_n \frac{\partial^2 u_n}{\partial t^2} = \nabla.\sigma_n + F_{vn} \tag{8}$$

where ρ is the density, kg/m^3; u is the displacement, mm; σ is the stress tensor; F_v is the external force per unit volume; $n = 1$, indicating the proppant, $n = 2$, indicating the rock matrix. The stress–strain relationship and strain–displacement relationship are shown in Eqs. (9) and (10):

$$\sigma = C.\varepsilon \tag{9}$$

$$\varepsilon = \frac{1}{2}\left[(\nabla u)^T + \nabla u\right] \tag{10}$$

where ε is the strain tensor; E is Young's modulus, GPa; v is the Poisson's ratio. As we assume that Young's modulus and Poisson's ratio of the reservoir rock would change from the outer surface of the fracture toward the inside as shown in **Figure 7** and the specific correlation is expressed as follows:

$$E = f(l) \tag{11}$$

$$v = f(l) \tag{12}$$

where l is the distance from the surface of the fracture toward the inside as shown in **Figure 7**. The closure pressure causes the proppant to embed into the fracture surface and the porosity of the propped fracture to change. The embedment of the proppant (h_{em}) can be expressed as follows:

$$h_{em} = u_m - u_p \tag{13}$$

where u_m is the displacement of the no-contact part between the fracture and the proppant under closure pressure; u_p is the displacement of the contact part between the fracture and the proppant.

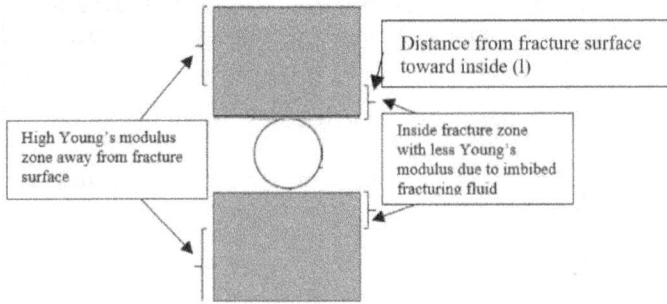

Figure 7.
Change in Young's modulus due to fracturing fluid interaction with the shale.

As the focus is proppant embedment into the shale formation, therefore, proppant is assumed as a rigid body. Thus, no deformation occurs in the proppant. However, the deformation of the shale surface is simulated with the penetration of proppant under uni-axial and bi-axial loads.

2.3 Calculation of fracture conductivity

CFD-CFX is used to calculate pressure drop and velocity across the inlets as shown in **Figure 3**. The finite volume method is adopted to solve the three-dimensional Navier–Stokes equations. Consistent with the experimental conditions for conductivity measurements, the flow was the steady state at 25°C. The continuity and momentum balance equations for the steady-state flow are shown below.
Continuity equation

$$\frac{\partial u}{\partial x} + \frac{\partial v}{\partial y} + \frac{\partial w}{\partial z} = 0 \tag{14}$$

Momentum equations

$$\rho\left(u\frac{\partial u}{\partial x} + v\frac{\partial u}{\partial y} - w\frac{\partial u}{\partial z}\right) = \frac{\partial P}{\partial x} + \mu\left(\frac{\partial^2 u}{\partial x^2} + \frac{\partial^2 u}{\partial y^2} + \frac{\partial^2 u}{\partial z^2}\right) \tag{15}$$

$$\rho\left(u\frac{\partial v}{\partial x} + v\frac{\partial v}{\partial y} - w\frac{\partial v}{\partial z}\right) = \frac{\partial P}{\partial y} + \mu\left(\frac{\partial^2 v}{\partial x^2} + \frac{\partial^2 v}{\partial y^2} + \frac{\partial^2 v}{\partial z^2}\right) \tag{16}$$

$$\rho\left(u\frac{\partial w}{\partial x} + v\frac{\partial w}{\partial y} - w\frac{\partial w}{\partial z}\right) = \frac{\partial P}{\partial z} + \mu\left(\frac{\partial^2 w}{\partial x^2} + \frac{\partial^2 w}{\partial y^2} + \frac{\partial^2 w}{\partial z^2}\right) \tag{17}$$

where x, y, z are the dimensions; u, v, w are the velocity directions in x, y, z directions; p is pressure inside fracture. Fracture permeability was determined according to Darcy's law provided in Eq. (18).

$$K = \frac{Q\mu L}{A\Delta P} \tag{18}$$

where K is the permeability, Q is the flow rate of the injected fracturing fluid, μ is the viscosity, L is the length of the fracture around the proppant, A is the cross-sectional area of the fracture zone, and ΔP is the differential pressure between inlet and outlet across the proppant. The conductivity of hydraulic fracture is generally defined as the maximum ability of the fracture to transmit a reservoir fluid through it. The conductivity is measured in $\mu m^2.cm$ based on the propped fracture width (cm) and permeability (μm^2).

$$\text{Fracture conductivity} = Kf * Wf \tag{19}$$

where Kf is the proppant permeability, μm^2 and Wf is the fracture width, cm.

3. Results and discussion

3.1 Experimental measured embedment

The main purpose of this experiment is to find the depth of the proppant embedded into the shale rock; the resulted embedment depths were used to validate the numerical model. **Figure 8** shows the images after the proppant embedment test with some embedment occurring on the shale core at the (a) bottom and (b) top of the shale core sample with the 20/40 mesh proppants, that is, proppant having a size between 20 and 40 μm. The embedment depths of proppant in shale formation soaked underwater are recorded in the experiments, which are 76 μm at the top and 64 μm at the bottom surface. **Table 3** shows the proppant embedment depth in soaked and unsoaked shale samples.

3.2 Numerical measured embedment

Contact analysis with finite element method has been performed to quantify the conductivity loss due to proppant embedment based on computational contact mechanics. A similar contact analysis has been also investigated between rail/wheel [26]. Initially, the model was developed and validated according to the boundary load as applied in the experimental procedure. As shown in **Figure 9**, the load is

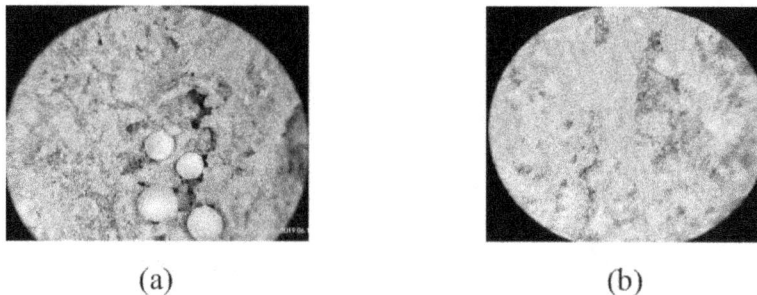

(a)   (b)

Figure 8.
SEM images of proppant (20/40 mesh) embedment on the original core soaked in fracking fluid: (a) bottom of the core and (b) top of the core.

No	Penetration with water (μm)	Penetration without water (μm)
(MAS 1)	305	170
(MAS 2)	310	150
(MSS)	290	170

Table 3.
Proppant embedment depth in the shale samples.

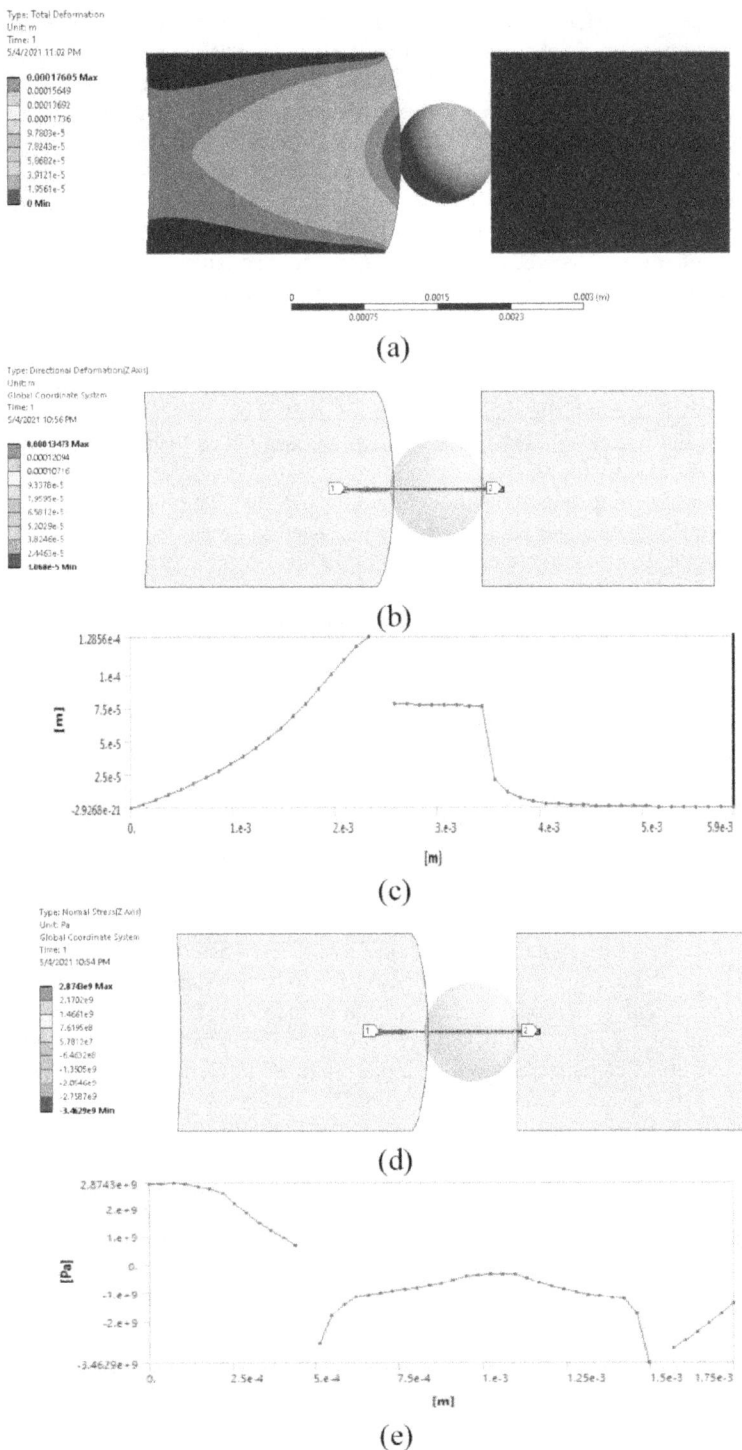

Figure 9.
Proppant under vertical principal stress: (a) total deformation and penetration due to an external load, (b) directional deformation along the Z axis, one surface under load while another surface in static, (c) plot of directional deformation profile, (d) normal stress profile along with the formation and proppant, and (e) plot of normal stress profile along with the formation and proppant under uni-axial load.

applied at the top surface, while the lower surface has been kept static to validate the results with a uni-axial compression test. In the case of the uniaxial load test, the resulted embedment depth in this study is found different for the top and bottom of the fracture surface, similar to an earlier experimental study performed using different types of proppants [39]. The significant difference between embedment profiles is the result of different proppant types being tested. 20/40 Ceramic proppants are rounder, more spherical, uniform, and stronger than 20/40 Ottawa proppants. This difference in proppants makes 20/40 Ottawa sand more prone to proppant embedment as well as any other damage mechanism caused by mechanical and chemical factors in fractures [39]. The present study shows that even though the proppant is strong enough to get deformed, and there is a penetration of proppant in the rock surface due to fracking fluid flow that has reduced Young's modulus of the fractured surface. Initially, ceramic proppant and rock properties are introduced based on the shale formation. A pair of contact between surfaces and proppant is defined. Then, the external loads are applied to the rock surfaces to obtain the stress transfer across the proppant of contact surfaces. The penalty-based method is used to simulate the contact behavior [27, 40]. The finite element method is a numerical method that can be successfully used to generate solutions for problems belonging to a vast array of engineering fields: stationary, transitory, linear, or nonlinear problems. For the linear case, computing the solution to the given problem is a straightforward process, and the displacements are obtained in a single step and all the other quantities are evaluated afterward. When faced with a nonlinear problem, in this case with a contact nonlinearity, one needs to account for the fact that the stiffness matrix of the systems varies with the loading, the force vs. stiffness relation being unknown before the beginning of the analysis. Modern software using the finite element method to solve contact problems usually approaches such problems *via* two basic theories that, although different in their approaches, lead to the desired solutions. One of the theories is known as the penalty function method [40]. The penalty method is simple to implement in practice. The penalty is a sort of friction coefficient, and one can specify a friction model that defines the force resisting the relative tangential motion of the surfaces in a mechanical contact analysis. By selecting a penalty, one can use a stiffness (penalty) method that permits some relative motion of the surfaces when they should be sticking. By applying the penalty method, the penetration of the proppant has been achieved higher at the top which is 75 μm, whereas the penetration at the bottom surface is recorded 60 μm. The penetration of proppant in numerical model has been achieved almost the same as recorded in the experiments, which was 76 μm at the top and 64 μm at the bottom surface. The similarity of the results shows that the developed model has satisfactory results and parametric study can be carried out for further analysis. Once the model has been validated with the experimental results, then the external force was applied on both sides of the proppant to represent the actual condition in the fracture formation.

3.3 Impact of embedment on fracture conductivity

In this section, the change in pressure and velocity of backflow of fluid across the proppant has been presented. Embedment has a profound impact on the pressure drop as well as velocity profile as shown in **Figure 10**. In this study, three different embedment cases have been considered (0, 60, and 80%). The percentage of embedment is defined as the proportion of the total proppant that is embedded through the fracture surface. Without embedment, a slight difference between inlet and outlet pressure has been recorded; however, a significant difference

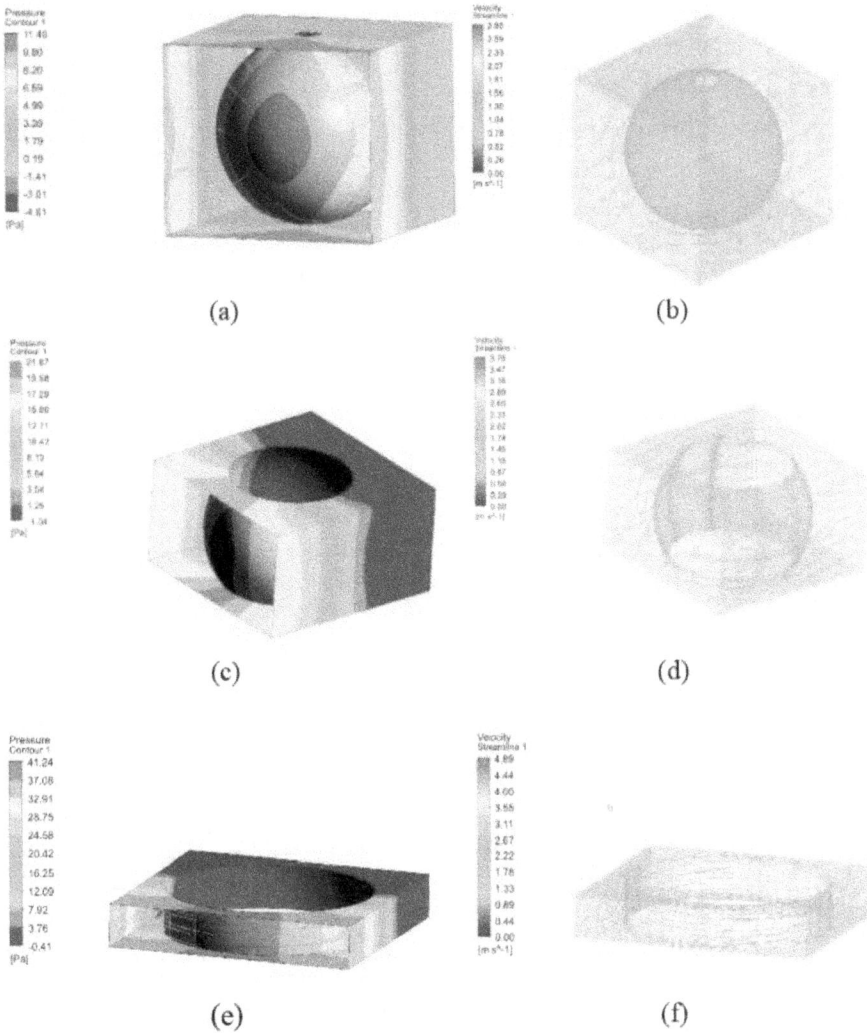

(a)

(b)

(c)

(d)

(e)

(f)

Figure 10.
*Velocity and pressure profiles in the fracture zone around the single proppant with and without embedment.
(a) Pressure profile with no embedment. (b) Velocity streamlines with no embedment. (c) Pressure profile with
60% embedment. (d) Velocity streamlines (a streamline is a line that is tangential to the instantaneous velocity
direction (velocity is a vector, and it has a magnitude and a direction. Color represents velocity magnitude) with
60% embedment). (e) Pressure profile with 80% embedment. (f) Velocity streamlines with 80% embedment.*

between inlet and outlet pressures can be seen at 60 and 80% embedment as shown
in **Figure 10(b)** and **(c)**. Inlet velocity in all cases is 0.5 m/s but around the
proppant, the flow velocity is recorded around 2 m/s and at outlet, the velocity is
achieved 1.5 m/s.

Based on the different embedment depths, the velocity of the injected fluid
varies significantly as shown in **Figure 11**. In all cases, injection velocity is constant,
that is, 0.5 m/s. A sudden increase in the fluid velocity is recorded around the
proppant and a decrease in velocities is presented at the end of the proppant.
The results show a significant decline in the velocity at 80% embedment;
therefore, fracture conductivity is recorded significantly low at high embedment
(see **Figure 12**). As fluid flowing continues around the exit sides of the proppant,
it begins to slow down due to eddy generated at the outlet/backside of the proppant.

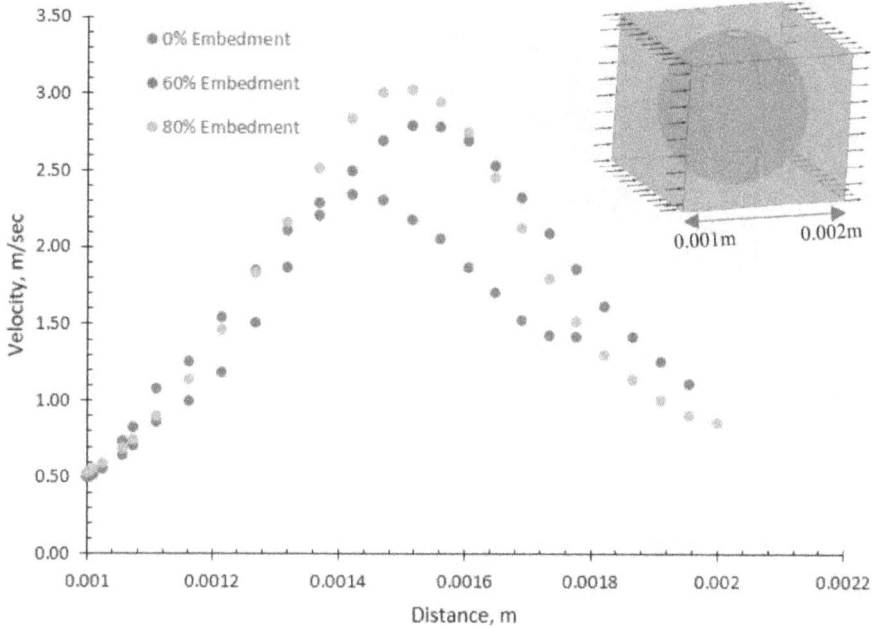

Figure 11.
Velocity profile of injection fluid around the proppant under different embedment percentages.

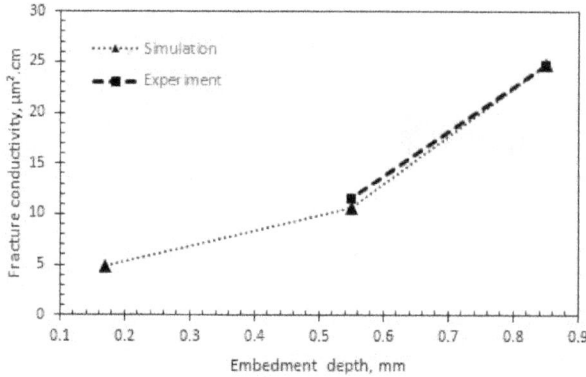

Figure 12.
Fracture conductivity obtained with finite volume method based on experimental and numerical measured embedment depths with finite element method.

The Lagrangian analysis is capable of revealing the underlying structure and complex phenomena in unsteady flows [41].

Distance is measured from one side (fracture inlet) of the proppant until the other side (fracture outlet) of the proppant. All the three positions of proppants have been presented in **Figure 10(a), (c),** and **(e)**. Numerical analysis is conducted by finite volume method to obtain pressure drop across the proppant and resultant fracture permeability. **Table 4** shows the fracture conductivity based on embedment percentage.

Finally, fracture conductivity is achieved based on fracture permeability and fracture width. **Figure 12** shows that fracture conductivity was measured based on

Embedment (%)	Fracture width (mm)	Height (mm)	Length (mm)	Injection flow rate, cc/min	Fracture conductivity (μm^2.cm)
0	0.85	1	1	17	26
60	0.34	1	1	17	9
80	0.17	1	1	17	5

Table 4.
Embedment and resultant fracture conductivity.

embedment depth obtained with experimental and finite element methods. A dramatic decrease in fracture conductivity has been obtained with the increase of embedment depth. The reason for a significant decrease in the conductivity is the significant pressure drop across the embedment. The results show that pressure loss at 60 and 80% embedment is 29,716 and 64,721 pa, respectively.

4. Conclusion

A numerical model is developed for contact analysis of proppant embedment in the formation based on experimental investigation. Initially, the model was developed based on the experimental design of proppant embedment in the laboratory where the load is applied uni-axially from the top. Then, the study is extended by applying load from top as well as from the bottom side of the proppant in the fractured surface to simulate the actual reservoir condition. The amount of proppant embedment has been computed on both sides of the proppant in the fracture surface. Also, the deformation and normal stress profile have been plotted along with the formation and proppant. The total penetration of the surfaces has been recorded 141 µm on each side as the equal loads have been applied on both sides of formations around the proppant. This shows that actual proppant embedment is very high if stresses are present on both sides of the proppant in the fracture. The computational contact mechanics analyses have been able to capture the actual conductivity of fracture showing that the finite element method can be used to estimate embedment depth and has comparable results with experimental measurements. Long-term production of hydrocarbon from shale reservoirs is directly related to fracture conductivity in the hydraulically stimulated reservoir volume. This study shows that the uncertainty and reduction in hydrocarbon production profile with time can be mimicked by exact estimation of proppant embedment and fracture closure with finite element method since it relates to fracture conductivity. The presented method can serve as a valuable criterion to effectively reduce the loss of hydraulic fracture conductivity in shale reservoirs with time. Based on this numerical model, the required fracture conductivity can be achieved by keeping the extra width of fracture in the design criteria to reduce the conductivity loss in the formation.

Acknowledgements

We acknowledge Petronas Research Fund (PRF) grant (0153AB-A33) awarded to E. Padmanabhan and Shale Gas Research Group, Universiti Teknologi Petronas for financial support to carry out this research work.

Conflict of interest

The authors declare no conflict of interest.

Author details

Javed Akbar Khan[1*], Eswaran Padmanabhan[1,2] and Izhar Ul Haq[1,2]

1 Shale Gas Research Group, Universiti Teknologi PETRONAS, Seri Iskandar, Malaysia

2 Department of Geoscience, Universiti Teknologi PETRONAS, Bandar Seri Iskandar, Malaysia

*Address all correspondence to: javedkhan_niazi@yahoo.com

IntechOpen

References

[1] Reinicke A. Mechanical and Hydraulic Aspects of Rock-Proppant Systems: Laboratory Experiments and Modelling Approaches. Potsdam: Deutsches GeoForschungsZentrum (GFZ); 2011

[2] Liu Y. Long term conductivity of narrow fractures filled with a proppant monolayer in shale gas reservoirs. Journal of Engineering Research. 2017;5(2)

[3] Song F, Qi F. Experiments of water's effect on mechanical properties of shale rocks. 2016

[4] Deglint H et al. Comparison of micro-and macro-wettability measurements and evaluation of micro-scale imbibition rates for unconventional reservoirs: Implications for modeling multi-phase flow at the micro-scale. Journal of Natural Gas Science and Engineering. 2019;62:38-67

[5] Fredd C et al. Experimental study of fracture conductivity for water-fracturing and conventional fracturing applications. SPE Journal. 2001;6(03): 288-298

[6] Man S. Compression and flow behavior of proppants in hydraulically induced fracture. Graduate Studies; 2016

[7] Ding X, Zhang F, Zhang G. Modelling of time-dependent proppant embedment and its influence on tight gas production. Journal of Natural Gas Science and Engineering. 2020;82:103519

[8] Bandara K, Ranjith P, Rathnaweera T. Improved understanding of proppant embedment behavior under reservoir conditions: A review study. Powder Technology. 2019;352:170-192

[9] Lyle D. Proppants Open Production Pathways. Artic: Schlumberger Ind; 2011. pp. 1-6

[10] Becerra M. Effect of overdisplacement of proppant in hydraulic fracturing treatments on the productivity of shale gas reservoirs. 2012

[11] Han J et al. Numerical study of proppant transport in complex fracture geometry. In: SPE Low Perm Symposium. Denver, Colorado, USA: Society of Petroleum Engineers; 2016

[12] Zhu H et al. DEM-CFD modeling of proppant pillar deformation and stability during the fracturing fluid flowback. Geofluids; 2018;**2018**:18

[13] Guo J et al. Analytical analysis of fracture conductivity for sparse distribution of proppant packs. Journal of Geophysics and Engineering. 2017; **14**(3):599-610

[14] Zhu H, Shen J, Zhang F. A fracture conductivity model for channel fracturing and its implementation with discrete element method. Journal of Petroleum Science and Engineering. 2019;**172**:149-161

[15] Li H et al. A new mathematical model to calculate sand-packed fracture conductivity. Journal of Natural Gas Science and Engineering. 2016;**35**: 567-582

[16] Ming C et al. Calculation method of proppant embedment depth in hydraulic fracturing. Petroleum Exploration and Development. 2018; **45**(1):159-166

[17] Pao W et al. Fill removal from horizontal wellbore using foam in different coiled tubing/annulus diameter ratios. International Journal of Oil, Gas and Coal Technology. 2015; **9**(2):129-147

[18] Khan JA, Pao WK. Fill removal with foam in horizontal well cleaning in

coiled tubing. Res. J. Appl. Sci. Eng. Technol. 2013, 6, 2655-2661

[19] Khan JA et al. Optimization of coiled tubing nozzle for sand removal from wellbore. Journal of Petroleum Exploration and Production Technology. 2020;**10**(1):53-66

[20] Khan J, Pao WK. Horizontal well cleanup operation using foam in different coiled tubing/annulus diameter ratios.American Journal of Applied Sciences. 2013;**14**:3235-3241

[21] Khan JA, Pao WK. Effect of different qualities of foam on fill particle transport in horizontal well cleanup operation using coiled tubing. In: Advanced Materials Research. Trans Tech Publications; 2014

[22] Khan JA et al. Comparison of machine learning classifiers for accurate prediction of real-time stuck pipe incidents. Energies. 2020;**13**(14):3683

[23] Khan JA et al. Quantitative analysis of blowout preventer flat time for well control operation: Value added data aimed at performance enhancement. Engineering Failure Analysis. 2020;**120**: 104982

[24] Shen Y et al. Impact of petrophysical properties on hydraulic fracturing and development in tight volcanic gas reservoirs. Geofluids. 2017;**2017**:5235140,13

[25] Dai C et al. Analysis of the influencing factors on the well performance in shale gas reservoir. Geofluids. 2017;**2017**, pp. 1-12

[26] Wang W, Shahvali M, Su Y. A semi-analytical fractal model for production from tight oil reservoirs with hydraulically fractured horizontal wells. Fuel. 2015;**158**:612-618

[27] Cooke C Jr. Conductivity of fracture proppants in multiple layers. Journal of Petroleum Technology. 1973;**25**(09): 1101-1107

[28] Akrad OM, Miskimins JL, Prasad M. The effects of fracturing fluids on shale rock mechanical properties and proppant embedment. In: SPE Annual Technical Conference and Exhibition, Denver, Colorado, USA. Society of Petroleum Engineers; 2011

[29] Zhang J et al. Experimental and numerical studies of reduced fracture conductivity due to proppant embedment in the shale reservoir. Journal of Petroleum Science and Engineering. 2015;**130**:37-45

[30] Fan M et al. Investigating the impact of proppant embedment and compaction on fracture conductivity using a continuum mechanics, DEM, and LBM coupled approach. In: 52nd US Rock Mechanics/Geomechanics Symposium. Seattle, Washington USA: American Rock Mechanics Association; 2018

[31] Wang W et al. Numerical simulation of fluid flow through fractal-based discrete fractured network. Energies. 2018;**11**(2):286

[32] Fjar E et al. Petroleum Related Rock Mechanics. Elsevier; 2008

[33] Li K, Gao Y, Lyu Y. New mathematical models for calculating proppant embedment and fracture conductivity. SPE J. 2015;**20**(2015):496-507.

[34] Khanna A et al. Conductivity of narrow fractures filled with a proppant monolayer. Journal of Petroleum Science and Engineering. 2012;**100**:9-13

[35] Andrei N. Penalty and augmented Lagrangian methods. In: Continuous Nonlinear Optimization for Engineering Applications in GAMS Technology. Springer.; 2017. pp. 185-201

[36] Nocedal J, Wright SJ. Penalty and augmented Lagrangian methods. In: Numerical Optimization. Springer: New York, NY, USA: Springer, 2006; pp. 497-528

[37] Wriggers P, Zavarise G. Computational contact mechanics. In: Encyclopedia of Computational Mechanics. 2004

[38] Xu J et al. Effect of proppant deformation and embedment on fracture conductivity after fracturing fluid loss. Journal of Natural Gas Science and Engineering. 2019;71:102986

[39] Corapcioglu H, Miskimins J, Prasad M. Fracturing fluid effects on Young's modulus and embedment in the Niobrara formation. In: SPE Annual Technical Conference and Exhibition. Amsterdam, The Netherlands: Society of Petroleum Engineers; 2014

[40] Stefancu A-I, Melenciuc S-C, Budescu M. Penalty based algorithms for frictional contact problems. Buletinul Institutului Politehnic din Iasi. Sectia Constructii, Arhitectura. 2011; 57(3):119

[41] Cagney N, Balabani S. Lagrangian structures and mixing in the wake of a streamwise oscillating cylinder. Physics of Fluids. 2016;28(4):045107

Review of Geochemical and Geo-Mechanical Impact of Clay-Fluid Interactions Relevant to Hydraulic Fracturing

Gabriel Adua Awejori and Mileva Radonjic

Abstract

Shale rocks are an integral part of petroleum systems. Though, originally viewed primarily as source and seal rocks, introduction of horizontal drilling and hydraulic fracturing technologies have essentially redefined the role of shale rocks in unconventional reservoirs. In the geological setting, the deposition, formation and transformation of sedimentary rocks are characterised by interactions between their clay components and formation fluids at subsurface elevated temperatures and pressures. The main driving forces in evolution of any sedimentary rock formation are geochemistry (chemistry of solids and fluids) and geomechanics (earth stresses). During oil and gas production, clay minerals are exposed to engineered fluids, which initiate further reactions with significant implications. Application of hydraulic fracturing in shale formations also means exposure and reaction between shale clay minerals and hydraulic fracturing fluids. This chapter presents an overview of currently available published literature on interactions between formation clay minerals and fluids in the subsurface. The overview is particularly focused on the geochemical and geomechanical impacts of interactions between formation clays and hydraulic fracturing fluids, with the goal to identify knowledge gaps and new research questions on the subject.

Keywords: Shale geochemistry and geomechanics, Clay minerals, Hydraulic fracturing fluids, Shale formation fluids

1. Introduction

Clay minerals interactions with fluids have gained attention in the petroleum industry because of their presence in source rocks, reservoir rocks and seal rocks in petroleum systems. In conventional reservoirs, interactions between clay minerals and fluids have been studied in relation to wellbore integrity and fines migration during production. The inception of enhanced oil and gas recovery, hydraulic fracturing and carbon storage technologies, highlighted knowledge gaps in terms of interactions between clays and fluids injected into the subsurface. Research efforts are focused to understand the impact of clay-fluid reactions geochemistry on shale geomechanics, and deciphering the mechanisms that drive these interactions in order to optimise various technologies adopted by the industry.

In retrospect, studies on interactions between clay minerals and formation fluids have been going on in the petroleum industry well before introduction of the advanced technologies alluded to above. These studies were mainly focused on the relationship between clay minerals interactions with formation fluids during formation, migration and deposition of hydrocarbons. For example, Drits et al. [1] studied clay mineral-fluid interactions in order to gain insight into transformation processes in clay minerals during generation and migration of hydrocarbons.

This review seeks to present a concise overview of published studies on interactions between clay minerals and various fluids in the subsurface with particular emphasis on hydraulic fracturing fluids. Reaction mechanisms as well as geochemical and geomechanical impacts are assessed.

2. History of clay-fluid interactions

Trends of research on interactions between formation clay minerals and fluids over the years have largely been determined by the exigencies of the petroleum industry. The drive for this is the need for in-depth understanding of reactions in order to characterise reservoirs and cap rocks, as close as possible to *in situ* conditions. Summary of research trajectories in major periods is explained in the following paragraphs and captured in **Figure 1**.

2.1 Petroleum formation and migration

The focus of researchers in the 1940s and 1950s, during worldwide oil and gas exploration, was to investigate the origins of petroleum. Reactions between clays and subsurface fluids were studied extensively. At that time, the major concern was assessment of quality of organic source rocks and the mechanisms involved in generation of oil and gas. In this regard, researchers such as Weaver [2, 3] and Sarkissian [4] recognised that analyses of clay rocks (shales) could be used to track the generation and migration of petroleum in source rocks. Weaver [2] noted that expandable clays are capable of withholding their pore water to greater depths. He therefore inferred that waters in expandable clays at greater depths were responsible for transporting hydrocarbons to reservoir rocks. This inference was premised on

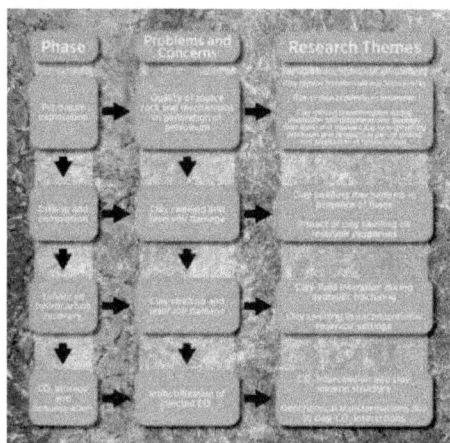

Figure 1.
Schematics of typical clay fluid interaction research topics and outcomes relevant to hydraulic fracturing of shale formations.

and supported by earlier studies where over 20,000 samples from major petroleum producing basins in the US showed strong statistical correlation between expandable clay minerals and hydrocarbon production. Similar to Weaver [2], Sarkissian [4] also studied petroleum deposits in the USSR and reported that clay minerals in argillaceous rocks were significant in the formation and deposition of petroleum resources. Premising on earlier works alluded to above, other researchers also used clay rock analyses to determine the hydrocarbon emplacement and migration times and for petroleum system analysis [1, 5–9]. Some of these works are summarised below.

Hamilton [5] used K-Ar dating to assess the formation of illite relative to the timing of generation and migration of hydrocarbons. Considerable correlation was found to exist between these two events. He reported that, in most cases, the timing of the expulsion of hydrocarbons was the same time authigenic illite formation ceased. He concluded that the link between clay-fluid interaction and petroleum generation and migration was therefore established, thus presence of authigenic illite could be used as an indicator of petroleum formation and migration.

Kelly [7] used mineralisation history present in fractures to reconstruct the migration history of hydrocarbons to their current reservoir and found that most petroleum migration paths showed preponderance of illite and clay mineral precipitates. He concluded that illite and other clay mineral precipitates can be used as an indication of petroleum migration pathway.

Jiang [6] examined clay minerals from the oil and gas perspective and drew a lot of parallels between various types of clays, their structural and geochemical transformations as a function of formation and transport of hydrocarbons. Jiang's work is different from earlier works in the sense that he investigates comprehensively the transformations that take place from deposition of rock to when petroleum is formed and expunged.

2.2 Drilling and completion

Adverse economic impact posed by swelling clays during drilling and completion caused intensification in studies of clay-fluid interactions with the aim of understanding the problem and solving it in the shortest possible time. Research was thus aimed at understanding the mechanisms that drive clay swelling during interactions with engineered muds. The conditions of clay swelling and accompanying complications were studied thoroughly with abundant literature to that effect [10–16].

Van Oort et al. [14] undertook an overview of the mechanisms guiding clay-fluid interactions. Their work identified the mechanisms by which various engineered drilling fluids suppress adverse reactions of clay minerals with water based fluids during drilling and completions. In concluding, the authors simplified their work by categorising drilling fluids into five groups based on the mechanism by which they stabilise clays in shale formations during drilling.

Shukla et al. [17] conducted a review of earlier works on clay mineral swelling in unconventional reservoir systems. They concentrated on the various conditions under which clays swell and the types of clay swelling. They also identified various clay stabilisers and gave a brief on how these work.

2.3 Enhanced hydrocarbon recovery

At inception of shale gas, tight sands and other unconventional petroleum systems development, enhanced hydrocarbon recovery techniques were at advanced stage. The need to fine-tune these technologies to the needs of unconventional reservoir systems spurred another era of research focussed on clay-fluid interactions in unconventional petroleum systems. One of the earliest works in this area was conducted by Zhou et al. [15] who premised their research on the fact that injected fluids caused formation damage due to clay swelling. They identified two types of swelling due to these interactions; crystalline swelling and osmotic swelling, the later posing significant adverse effects on reservoir quality. Alalli et al. [18] also noted that injected fluids caused disequilibrium in formation which leads to dissolution and precipitation of minerals in an attempt to return to equilibrium state. Dissolution and precipitation patterns were thus examined in order to identify their impact on reservoir quality. Dissolution of minerals, they noted, enhanced the porosity and permeability of formation by creating additional pore volume and linking previously unconnected pores; whereas precipitation of new minerals had adverse impact on reservoir quality due to the occlusion of flow paths, due to mineral growth within the existing pore space.

Buller et al. [19] analysed the Haynesville shale play in East Texas to understand what factors were responsible for efficiency of hydraulic fracturing in this formation. Their work concluded that, in high clay content zones, the efficiency of the fracturing was low due to massive proppant embedment and migration of fines. They postulated that post fracturing diagenetic events could also be initiated due to clay minerals interaction with fracturing fluids.

A similar effort was undertaken by Radonjic et al. [20] in their research focused on Caney shale. They sought to draw the link between mineralogical composition and microstructure of Caney shale to mechanical responses in order to delineate formations suitable for fracturing as well as predict mechanical responses of these formations.

2.4 Geological CO_2 storage

Recent surge in research on interaction between clay minerals and CO_2 in the subsurface is due to the advent of the concept of CO_2 capture, utilisation and storage (CCUS). The CCUS technology may also incorporates enhanced hydrocarbon recovery when CO2 injection is done in depleted oil and gas reservoirs. The importance of clay-fluid reactions is seen in the fact that most target storage reservoirs and accompanying seal formations have high clay mineral contents. In addition, reactions that cause immobilisation of CO_2 and the ability of caprock to withstand pressures resulting from CO_2 plume beneath are directly related to the amounts and types of clay minerals in the formation [21, 22]. Many studies have thus been conducted to investigate clay-fluid interaction in the context of understanding their implications on combined enhanced hydrocarbon recovery and CO_2 storage projects [23–29].

Olabode and Radonjic [27] studied the reaction between CO2-Brine and caprock formation to assess the impact of mineral precipitation patterns on caprock integrity at elevated temperature and pressure conditions. They concluded that, precipitation of minerals could cause the sealing of micro-pores in caprocks thus enhancing their ability store CO2 within the subsurface. Hui Du et al. [30] also studied the sealing properties of caprock at nano and micro-scale with the premise that the durability of the caprock is directly affected by the nanostructures and microstructure of these rocks.

3. Clay minerals

3.1 Chemical composition & crystallographic structure of clay minerals

Clay minerals are a product of rock weathering, and form from decomposition of feldspar minerals in hard rocks such as granite. They are commonly described as soil particles with sizes below 2 μm, often labelled as nature's nanoparticles. In terms of chemical composition, clays belong to a group of minerals called alumino-silicates. The alumino-silicates are composed of complex arrangement of atoms to form diverse structural configurations with the basic components being silicon, aluminium and oxygen. Silicon and aluminium atoms bond with oxygen to form silicon tetrahedral sheets and aluminium octahedral sheets respectively. These sheets are subsequently bonded by sharing common oxygen atoms, though the oxygen atoms at the edges of both sheets are left unpaired. These unpaired oxygen atoms at the edges of the sheets impose negative charges on clay mineral surfaces rendering them water sensitive and highly reactive to cations [31, 32].

Another factor contributing to high negative charges in clay minerals is the isomorphic cationic substitution within the sandwiched tetrahedral and octahedral sheets, which leads to imposition of excess negative charges on clay mineral surfaces [33]. The mechanism described above contributes to higher levels of clay sensitivity to water-based engineered fluids in subsurface.

3.2 Classification of clay minerals

Hughes [34] was one of the earliest researchers to attempt classification of clay minerals relative in the petroleum industry. This classification was done a few years after commercialization of X-Ray Diffraction (XRD) technology, which was hitherto used by petroleum companies as a method of studying clay minerals. Hughes [34] classified clay minerals into: Kaolinite, Smectite, Illite and Chlorite groups as shown on **Table 1**. He also indicated other classes which are mainly mixed layers of the four groups of clays. Descriptions by Hughes [34] are captured below:

- Kaolinite is composed of one silicate tetrahedral and aluminium octahedral thus a 1:1 clay mineral. This structure makes kaolinite relatively stable due to its low surface area and adsorption capacity.

- Smectites are composed of two silicate tetrahedrals bonded with one aluminium octahedral thus a 2:1 clay mineral. Smectites have a very high rate of expansion and/or shrinkage and are by far the most problematic clay minerals during drilling and production especially with water-base engineered fluids. This behaviour is attributable to the large surface area and high cation exchange capacities of smectites consequently leading to high adsorption capacity.

- Illite is composed of tetrahedral and octahedral plates arranged in a 2:1 format just like smectites. They however have lower adsorption capacities than smectites but higher than kaolinites.

- Chlorites consist of Brucite layers alternating with three-sheet pyrophyllite type layers. Though Chlorite may occur as macroscopic or microscopic crystal, they often occur as mixtures with other minerals in the microscopic state.

3.3 Clay mineral properties

Clay minerals are unique in a number of properties they exhibit; however, the following attributes of clay minerals have significant impacts on their interactions with fluids and are briefly captured below.

Clay Type	Chemical Formulae	Surface Area (m^2/gm)	CEC $(meq/100\ g)$	Configuration
Kaolinite	$Al_4[Si_4O_{10}](OH)_8$	20	3–15	1:1
Illite	$(K_{1-1.5}Al_4[Si_{7-6.5}Al1-_{1.5}O_{20}](OH)_4)$	100	15–40	2:1 Non-Expandable
Smectite	$(0.5Ca, Na)_{0.7}(Al, Mg, Fe)_4[(Si, Al)_8O_{20}]{\bullet}nH_2O$	700	80–100	2:1 Expandable
Chlorite	$(Mg, Al, Fe)_{12}[(Si, Al)_8O_{20}](OH)_{16}$	100	15–40	2:1 Non Expandable

Table 1.
Four major types of clay minerals relevant to hydraulic fracturing.

3.3.1 Cation exchange capacity (CEC)

Cation exchange capacity (CEC) is defined as the amount of positive ion substitution that takes place per unit weight of dry rock [35] and is expressed in meq/100 g (milliequivalents per one hundred grams) of dry rock. Substitution of ions in minerals is the product of interfacial electrochemical interactions. Some of the most common cations exchanged are calcium (Ca^{2+}), magnesium (Mg^{2+}), potassium (K^+), sodium (Na^+) and ammonium (NH_4^+). CEC controls contribution of clay minerals and clay-bound water to electrical conductivity of rocks as well as the wettability characteristics of clay minerals during clay-fluid interactions.

Researchers developed various methods of measuring CEC over the years with more accurate methods still being developed. Some of the earlier methods have been exhaustively discussed in literature [36–40]. The most common methods currently used for CEC determination include: wet chemistry method; multiple salinity method and membrane potential method. These are however not without their limitations.

Bush and Jenkins [40] developed a method based on the use of the wet chemistry method in which several samples were investigated and a plot of best fit generated. The main challenge with their method is that, some minerals are capable of adsorbing water in humid environments though they have no CEC. Bush and Jenkins [40] proposed their method as a supplementary method for the wet chemistry method rather than a replacement.

Cheng and Heidari, [41, 42] introduced a new theoretical model of measuring CEC based on energy balance between chemical potential and electric potential energy. This involved the combined analysis of data collected from XRD (X-Ray Diffraction), NMR (Nuclear Magnetic Resonance) and nitrogen adsorption–desorption isotherm measurements with direct evaluation of CEC based on ammonium acetate method and Inductively Coupled Plasma Mass Spectrometry (ICP-MS) measurements used in cross-validation of the results. They however alluded to the fact that their method was yet to be developed for complex rock composition.

3.3.2 Clay swelling

Clay swelling results mainly from fluid intake into the inter-layered structure of clay minerals. Electrochemical interactions between clay minerals and fluids are central to the swelling of clays. The type, quantity and charge of cations in the interlayer zones of clay are the main driving forces in the swelling process. Clay swelling and formation damage during enhanced oil recovery have also been discussed extensively [43, 44].

Two main types of swelling mechanisms have been identified in clay minerals which include crystalline swelling and osmotic swelling [45, 46]. During crystalline or surface hydration mechanism, the water molecules are adsorbed on the crystal surfaces with hydrogen bonding holding the water molecules to oxygen atoms exposed from the crystal surface. Subsequent layers of water molecules align to form a quasi-crystalline structure between unit layers, which results in an increased c-spacing. This type of swelling is common to all types of clay minerals, although to a different degree. In osmotic swelling mechanism, the concentration of cations between unit layers in clay minerals is higher than that in the surrounding water, water is therefore osmotically drawn between the unit layers and the c-spacing is increased. Osmotic swelling mechanism causes a larger swelling relative to the crystalline swelling but only a few clay minerals, such as sodium montmorillonite, swell in this manner [47].

4. Hydraulic fracturing

Hydraulic fracturing entails high rate injection of pressurised fracturing fluids into low permeability formations often targeted at specific horizontal sections of a wellbore in order to induce failure, consequently fracturing rock formation and creating a fracture network that can provide permeability in otherwise almost-impermeable rocks. Studies have shown that the fractures induced by hydraulic fracturing fluids are formed normal to the direction of minimum horizontal stress in the horizontal section of the wellbore. Horizontal wells are normally drilled in trajectories parallel to the minimum horizontal stress in a given reservoir. However, branch-like networks of micro-fractures are formed in all directions, resulting in a hydraulic connectivity that provides permeability form otherwise imperme-able shale matrix. The majority of fractures are kept open by proppants which are transported by the injected fluid into the formation. Proppants ensure that frac-tures remain open thus enhancing the contact area between reservoir and wellbore which consequently serve as a conduit for hydrocarbon recovery, from otherwise low permeability shales [48, 49].

Hydraulic fracturing entails lots of activities, thus, research is fine-tuned on investigating and understanding certain key issues about hydraulic fracturing. For example, Rikards et al. [50] indicated that one of the biggest problems in hydraulic fracturing has to do with ability to find balance between proppant-quality and proppant-transport efficiency. They intimated that high density proppants pose proppant transport challenges whilst low density proppants present issues of strength of the proppants. Also, the importance of fluid viscos-ity in terms of providing sufficient fracture width to enable transport and proper placement of proppants is another issue in hydraulic fracturing highlighted by Montgomery et al. [51].

4.1 Hydraulic fracturing fluids (HFF)

Since inception of the concept of hydraulic fracturing, a lot of fluids have been developed and experimented as possible suites for various formation types and even geographical locations. These are discussed below.

4.1.1 Water-based fracturing fluid

Water-based fracturing fluids are the most common hydraulic fracturing fluids in use today. This is due to their low cost, availability and their ability to transport proppants in place to maintain fracture conductivity. Though water-based hydraulic fluids have several advantages over other types of fracturing fluids, they are more susceptible to causing formation damage due to hydration of clays which may lead to lower recovery rates for hydrocarbons. Ribeiro and Sharma [52] contend that water-based fracturing in unconventional wells, most of which contain substantial clay mineral component, presents significant challenges. One of the most effective ways of dealing with this drawback, thus, has been to use energised water-based fracturing fluids in which the fracturing fluid is energised with CO_2 or N_2. This significantly reduces the amount of water needed for fracturing and thus improves the fracturing job in water-sensitive formations. Some water-based fracturing fluid types are discussed below.

Slickwater fracturing fluids are primarily composed of water, sand proppants and other chemicals to deal with friction, corrosion, clay swelling and other adverse reactions due to injection of fluids into the subsurface. These fluids are characterised by lower viscosities and the ability to generate complex fractures which generally reach deeper into target formations. The drawback with this type of water-based fracturing fluid is its poor proppant transport capacity. This is often compensated for with higher pumping rates in order to maintain optimal velocities that prevent settling of proppants.

Linear fracturing fluids were developed as a solution to the poor proppant carrying capacity of slickwater fluids. This was achieved by increasing the viscosity of fracturing fluid through addition of polymers in the fluids. These polymers are capable of turning the aqueous solutions into viscous gels capable of transporting proppants effectively but may also adversely affect the permeability of low permeability formations by forming filter cakes on the walls of fractures. Linear fracturing fluids are good in controlling fluid loss in low permeability formations but prone to higher fluid losses in high permeability formations.

Cross-linked fluids were developed to obtain increased viscosity and performance of gelled polymers without necessarily increasing the concentration of polymers. To develop these fluids, Aluminium, Borate, Titanium and Zirconium compounds may be used to crosslink hydrated polymers in order to increase the viscosity of resulting fluid. The main advantage of these fluids is the reversibility of crosslinks based on pH adjustments. This enables better clean up and consequently improved permeability following fracturing treatment. Borate crosslinked fracturing fluids have been reported to show rheological stability, good clean-up and low fluid loss up to temperatures of over 300°F.

In viscoelastic surfactant gel fluids, increased viscosity and elasticity is obtained by adding surfactants and inorganic salts into water-based fracturing fluids to create ordered structures. These fluids exhibit very high zero-shear viscosity and are capable of transporting proppants with lower loading and without the comparable viscosity requirements of conventional fluids [53].

4.1.2 Oil-based fracturing fluid

Oil-based fracturing fluids are used mostly where a formation is water-sensitive perhaps due to the presence of large quantities of expandable clay minerals. Oil-based fracturing fluids have been found to better preserve fracture conductivity [54, 55] as well as provide better performance in terms of proppant transport due to the generally higher viscosity and lower specific gravity. In their work on wells located in Bakhilov and North Khokhryakov fields in Western Siberia, Russian Federation, Cikes et al. [54] studied the responses from wells after treatment with oil-based fracturing fluids in a depleted oil reservoir. These wells were initially fractured with water-based fracturing fluids but the treatment failed and did not yield significant improvements in the productivity especially for the long term. Following fracturing with oil-based fracturing fluids, over ten-fold production improvement was witnessed relative to pre-fracturing productivity.

Another advantage of oil-based fracturing fluids as noted by Hlidek et al. [56] is that they are easier to clean-up and can be re-used. Hlidek et al. [56] compared the cost of using water-based fracturing fluids to oil-based fracturing fluids in the Montney (Canada) unconventional gas development. Based on their comprehensive analysis, they concluded that the cost of using oil-based fluids was lower in the long term since all the load oil could be recovered within 4 to 8 weeks and could be reused in fracturing. The main disadvantage of oil based fracturing fluids however, is environmental damage when not properly disposed of.

To enhance the efficiency and recovery of oil-based fracturing fluids, CO_2 has been employed in energising these fluids. Energising oil-based fluids significantly reduces the amount of fluid required to fracture a specific formation as well as aids in fluid recovery following the fracturing process [57, 58]. Vezza et al. [58] studied the impact of energised oil-based fracturing fluid in Morrow Formation in Southern Oklahoma where they used gelled diesel/CO_2 as fracturing fluid. Their results indicated an overall increase in production rate and predicted long-term stability of the wells. Gupta et al. [57] also reported improvements in well productivity and stability after using energised gelled hydrocarbons in fracturing treatments. In their study, they compared the use of conventional gelled fluids to CO_2 energised gelled fluids in formations in Canada. Their conclusions were that: The use of energised gelled hydrocarbon fracturing fluids led to improved production relative to conventional gelled hydrocarbon fluids.

4.1.3 Gas fracturing fluid

Gas fracturing involves the injection of gas at high pressures into the subsurface in order to create fractures within targeted reservoir locations. Nitrogen gas is the most employed gas for fracturing purposes, due to its obvious advantages of availability, its inert nature and of course, cost [59]. The main limitation of gas fracturing is the depth it can be used as a fracturing fluid since it has a low density and thus is restricted to reservoirs of less than 5000 feet deep [60]. Recent advancements in ultra-light weight proppants [50, 61] provides positive prospects that may counter the depth limitation of gas fracturing to some extent.

4.1.4 Foam-based fracturing fluid

Another type of fracturing fluid is a foam-based fracturing fluid which is generated from the combination of two phases of liquid and gas as well as addition of surfactant to ensure stability [51]. The main advantages of this type of fracturing fluid is its efficiency in water-sensitive areas and its relatively better proppant carrying capacity compared to water-based fracturing fluids [62]. High cost and risk of flammability are the main disadvantages of foam-based fracturing fluids.

4.1.5 CO_2-based fracturing fluid

Consideration of the use of CO2 as a fracturing fluid was mooted due to problems encountered with water-based fracturing fluids in terms of permeability damage. Liquid CO2 is considered an alternative to water-based fracturing fluids because it causes minimal formation damage plus clean-up is easily achieved. Lower viscosity, miscibility of CO2 with hydrocarbons, ease of displacing methane from organic matter and the ease of recovery of CO2 enables it to create extensive and complex fractures at lower breakdown pressures [63, 64]. However, high pumping rates needed to enhance proppant carrying capacity of CO_2 makes CO_2-based fracturing fluids relatively expensive. Additionally CO_2 is not readily available at all sites. The future applications may change if CO_2 can be sequestered.

5. Mechanisms of clay minerals-HFF interactions

5.1 Hydration imbibition and fluid retention

Imbibition describes the displacement of immiscible fluids from within the formation matrix. In the context of this topic, the fluid within the formation matrix is hydrocarbon and brine, whereas the invading fluid is the hydraulic fracturing fluid, mostly water. The displacement described above occurs at times where the fracturing fluid comes into contact with the formation face creating disequilibrium. In order to gain equilibrium, fracturing fluid is drawn into the matrix spontaneously, without the application of any form of pressure. This phenomenon is known as spontaneous imbibition. Handy [65] defined it as the process in which a fluid is displaced by another fluid within a porous medium due to the effect of capillary forces alone. Other researchers like Bear [66], Bennion et al. [67], Hoffman [68] and Dutta [69] have also interrogated the mechanisms of hydration imbibition.

Imbibition of water into shale matrix has been identified as the major water retention mechanism when using water-based hydraulic fracturing fluids [70, 71]. Research into the controlling factors of imbibition of water in fracturing fluid into formation matrix revealed it to be the function of several parameters which are briefly discussed in the following:

5.1.1 Fluid and rock properties

Fluid and rock properties have been identified as determinants of the amount and rates of imbibition. Ma et al. [72] reported that when water displaces oil and gas within the formation, the rate of imbibition becomes directly proportional to the viscosity of water. Pore sizes of formation inversely affect imbibition since smaller pore sizes generate greater capillary pressures and thus higher imbibition.

5.1.2 Initial water saturation

The initial amount of water present in matrix of rocks has been investigated by several researchers to ascertain its impact on the quantum of imbibition, but findings have been inconsistent, thus making it difficult to draw any conclusions. Whereas Blair [73] and Li et al. [74] found that an initially high water saturation of formation led to lower volumes of imbibed water, Cil et al. [75] and Zhou et al. [76] found the opposite in their experiments. Other works by Li et al. [74], Viksund et al. [77] and Akin et al. [78] also concluded that initial water saturation had no effect on imbibition of water by the formation. They explained that volume of water imbibed is a function of capillary pressure and effective permeability but these show an inverse and direct relation with water saturation respectively. The amount of water imbibed is therefore not controlled by a single parameter, but will depend on which of the two variables is dominant in any given formation. In this regard, they concluded that the influence of initial saturation on imbibition should be ascertained for every formation independently.

5.1.3 Temperature

The impact of temperature on imbibition is not direct however; temperature of a formation has impact on wettability and fluid properties which subsequently impact imbibition. Experimental investigations by Handy [65], Pooladi-Darvish and Firoozabadi [79] concluded that higher temperatures led to faster rates of imbibition.

5.1.4 Clay content

Total clay content directly relates to the effect of pore size on rate and amount of imbibition. Due to small pore size of clay-rich rocks, higher clay content in a formation results in smaller pore sizes thus greater imbibition. This position is confirmed by Zhou et al. [80] who performed several experimental and numerical analyses on the Horn Shale gas formation and concluded that high clay content in a formation leads to high volumes and rates of imbibition respectively.

5.2 Osmosis

According to Zhou et al. [81], earlier researchers viewed imbibition to mainly be the product of capillary pressure but findings from recent studies have challenged this position. Recent research shows osmosis contributes significantly to water imbibition and thus clay minerals and hydraulic fluid interactions especially for unconventional reservoirs which are often characterised by high clay mineral contents.

During osmotic imbibition of water into formation, formation clay minerals act as semi-permeable membranes through which fracturing fluids invade the matrix of the formation. Here, solutes from the concentrated formation fluids try to move into lower solute fracturing fluids but due to the semi permeable membrane formed by presence of clay, the solutes are unable to cross this barrier. Continuous accumulation of solutes near the semi-permeable membrane creates an attraction force that draws water into the formation in order to balance out the concentration differential.

6. Geochemical and geomechanical impacts of clay mineral-hydraulic fracturing fluids interaction

6.1 Water-blocking effect

Inorganic clayey matrix is generally known to be water-wet therefore providing favourable conditions for imbibition of water from fracturing fluid within fractures. In this process, the invading water displaces gas from the surface of clay matrix which leads to the formation of a multiphase flow environment near the fracture surface (**Figure 2**). Development of this phenomenon can create an unfavourable saturation condition, under which gas flow through fractures is hindered, thus lowering yield for wells. The phenomenon is known as water-blocking and it is has been described by researchers as one of the most severe damages in reservoirs with ultra-low permeability [82–84].

Recent experimental works on the imbibition of water by shale rocks showed that the imbibed water remains within the pore network, thus reducing the permeability to gas of the reservoir [81]. Simulation and history matching also confirmed that invasion and wetting of clay mineral surfaces by water from fracturing fluid

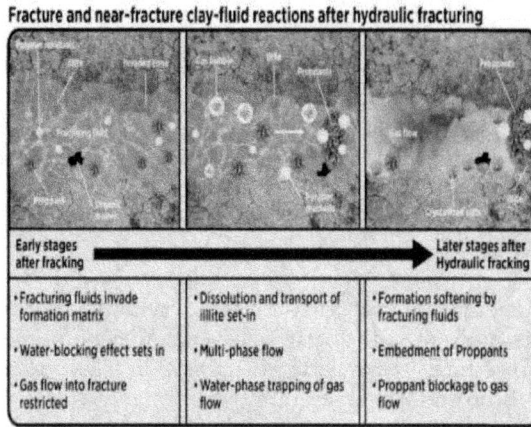

Figure 2.
Fracture and near-fracture clay-fluid interactions (adapted from [103]).

was responsible for decline in gas production. Reduction in gas flow due to water blocking effect has also been reported by Shanley et al. [85], who observed drastic reduction in gas production when water concentrations in fractures exceeded 40–50%. Detailed study of water-blocking phenomenon has showed that this phenomenon may cause permanent damage for some shale formations whiles the damage is transient for other shale types [70, 71, 86]. The details of mechanisms and variables that determine whether damage is temporal or permanent are still being investigated.

6.1.1 Water-blocking effect as a transient effect

Water-blocking during fracturing of unconventional reservoirs is explained by the presence of two pore types in unconventional formations. The first pore types are the larger oil-wet pores located within the organic matrix of the formation. The second type of pores are the smaller water-wet pores located within the inorganic argillitic matrix. Pore throats of the larger oil-wet pores are however small. During hydraulic fracturing, high pressure fluids break the formation to form fractures with some fracturing fluids leaking off into near-fracture matrix. Once in the matrix, the fluids first occupy the larger oil-wet pores. However, due to smaller pore throats, fracturing fluid in the formation is segmented within each internal pore with minimal linkage to other pores. This causes water to be domiciled in formation as droplets filling larger oil-wet pores which subsequently makes remobilisation difficult upon resumption of production. This phenomenon significantly reduces hydrocarbon effective permeability. The natural healing process in this phenomenon occurs when fluid is drawn from larger pores into smaller water-wet pores deeper within the reservoir thus dissipating the water blocking effect. This leads to improved permeability and hydrocarbon production [71].

6.2 Mineral dissolution and precipitation

Clay minerals and non-clay minerals (carbonates and quartz) within a formation are susceptible to geochemical attack from the fracturing fluids. Most shale formations were deposited in sea water-rich environments and have established equilibrium of their minerals and fluids over geological time. Once these formations are exposed to engineering fluids, especially water-based fluids, the geochemical equilibrium is no longer stable. Subsurface temperature, pressure, and pH often enhance geochemical reactivity of scale-forming minerals, resulting in changed porosity, and fracture permeability as a result of mineral dissolution and precipitation [87, 88].

Dissolution of rock forming minerals has been reported at low pH. As pH increases, ions from dissolved minerals recrystallise to form new minerals and/or amorphous precipitate that may have an adverse effect on formation permeability. At very high pH, clay minerals within a formation become unstable and may become mobile. This situation leads to migration of illite samples which may occlude the hydrocarbon flow paths within the formation (**Figure 2**).

6.3 Shale swelling

Shale swelling during hydraulic fracturing results from swelling of clay minerals within fracture face and shale matrix (**Figure 2**). Three mechanisms have been noted to cause clay mineral swelling as water is adsorbed into nano-pores, micro-pores, meso-pores and even macro-pores of clay minerals in the formation. The first mechanism is swelling due to hydration of negatively charged clay surfaces with several water layers depending on the type of clay. This has been observed in various types of clays with different levels of water saturation [89, 90]. The second mode of swelling is similar to what causes imbibition of water into clay minerals, where the clay acts as a semi-permeable membrane. In this case water is moved into the inter-layer spaces of clay minerals causing massive swelling [91]. In the third mechanism, continuous expansion of clay interlayer leads to separation of the clay layers into different clay components thus transforming initially intact clay layers into inter-particle spaces [92, 93].

Swelling of clays within the fracture walls can lead to constricted apertures which severely restrict flow of hydrocarbons during production [94]. Clay swelling may also induce micro fractures in formations which may improve absolute permeability [95, 96].

6.4 Stress development

During fracturing, interactions between shale and invading fluids lead to swelling of clay minerals within the shale causing in-situ stress development. Osmosis has been suggested as the potential transport mechanism which causes swelling pressure build-up within shale rock. The reaction between clay minerals and invading fluid is observed to be a primary cause of damage to reservoir permeability during and after hydraulic fracturing. Previous experiments have shown that a chance of permeability impairment following fluid interaction with formation is directly related to the specific clay mineral content of the formation [31, 71, 97]. In the case that the solute concentration in fracturing fluid is significantly lower compared to concentrations in clay interlayer, osmotic swelling is likely to occur, where fluid is drawn into the interlayer with the aim of balancing the solute concentrations. This phenomenon leads to significant expansion of the clay minerals. Expansion of clay minerals therefore exerts pressure on surrounding pores and matrices thus leading to a build-up of stress.

6.5 Mechanical weakening

Geo-Materials mechanical properties are dictated by the amount of pore space present, compositional heterogeneities [98], solids (inorganic and organic) mechanical strength and the presence/absence of pore-fluids and their composition. Therefore, mechanical weakening of formation rocks due to reaction with newly introduced fracturing fluids has been observed by a number of researchers. Akrad et al. [99] observed that sustained interaction between fracturing fluid and formation can induce softening of the formation rock thus reducing the Young's Modulus of the rock, therefore producing mechanical weakening. Du et al. [98] investigated mechanisms of fracture propagation due to hydraulic fluid injection and concluded that the mechanical response of formations due to interaction with fracturing fluid is directly linked to the mineral composition and geochemistry of rocks. In their studies of proppant embedment efficiency, Corapcioglu et al. [100] found that exposure of the formation to fracturing fluid leads to decrease in Young's Modulus of the rock. Research on the impact of fracturing fluid on formation mineralogical components has also showed that non-clay minerals (carbonates, quartz, feldspar and various sulfides/sulphates) as well as clay minerals, like chlorite, are susceptible to dissolution in fracturing fluids, leading to a reduction in the structural strength of the formation.

LaFollette and Carman [101] reacted Haynesville shale samples in fracturing fluid at temperature of about 300°F for periods of 30, 60, 120 and 240 days respectively and observed changes in Brinell Hardness of the samples. The highest reduction occurred between 60 to 120 days after which there was a marginal increase in hardness. Carman and Lant [102] also reacted rock samples with different fracturing fluids at temperatures close to subsurface formation temperatures. Their results showed that Brinell's Hardness for all the rock samples decreased after reacting with fracturing fluids.

6.6 Impact of geochemistry of shale-HFF reactions on toxicity of produced water

Flowback water is the fluid produced immediately following treatment of a well. This fluid is generally made up of mixed compositions of fracturing fluids, products of reactions between fracturing fluids and formation, and formation fluids. High salinity and concentrations of dissolved metals have been reported in flowback waters [104, 105]. Potential toxicity of produced water to humans and the environment remains high and of great concern; therefore, researchers have been studying these fluids to assess their risks to the environment. Most studies of this nature have largely been undertaken in producing unconventional hydrocarbon fields in the USA as well as black shales from Germany and Denmark [106, 107]. A summary of these works is presented below:

Chapman et al. [104] sought to characterise the flowback waters from Marcellus Formation in the Appalachian Basin with strontium isotopes in order to help detect and trace contamination of surface and ground waters by flowback water. The geochemical characterisation of the flowback fluids showed elevated levels of Bromide, Calcium, Strontium (up to 5200 mg/L), Sodium, Chloride (up to 12000 mg/L) and Barium. They also reported high total dissolved solids in the fluids in excess of 200000 mg/L. They concluded that high elemental concentrations were the result of interactions between fracturing fluids, formation minerals and formation fluids.

For their part, Wilke et al. [107] studied the rate of release of metals, salts anions and organic compounds from shale rocks in Denmark and Germany. They reported that, concentration of ions in solution largely depended on the composition of the black shale and did not show dependency on pH of the fluids used in experiments. There was however a correlation between the buffering capacity of specific mineral components of the rock, such as pyrite and carbonate on amount of dissolved ions. They also reported decrease in ionic concentrations over time due to precipitation of new minerals. Findings from this research provided an understanding of possible flowback water composition and toxicity from these unconventional fields in Denmark and Germany respectively.

Experimental studies by Macron et al. [106] on geochemical interactions between fracturing fluid and formation showed evidence of clay and carbonate dissolution as well as precipitation of new minerals. Dissolution leads to elevated elemental concentrations in fluids some of which show a drastic reduction when precipitation of new minerals begins. Results from this experimental work were validated using measured concentrations of sodium chloride (NaCl) in Marcellus shale which showed similar trends. They concluded that results from their work can be used to form a basis of assessing controls of geochemical reactions in the reservoir, in other words, flowback water compositions.

7. Summary of advances and research gaps topics for future consideration

In this section, recent research advancements are summarised. Review of recent studies shows that geochemical and geo-mechanical impacts from interactions between clay minerals and/or formation and fracturing fluids are being assessed more closely to help solve problems associated with reduced permeabilities during post-fracturing flowback. Geo-mechanical response of formations due to differences in temperatures of fracturing fluid and subsurface formations have also become a focus area for researchers. These advancements have unlocked new areas of research which will be explored in the near future. Other researchers have also focused on developing methods to measure the extent to which geochemical and geo-mechanical impacts are controlled by certain mechanisms during interactions between formations and fracturing fluid. Recent studies assessed for the purpose of this review are as follows:

7.1 Fracture face damage

Though water-blocking effect is known to be one of the causes of permeability loss following hydraulic fracturing, mechanisms by which this occurs are not well understood. Elputranto et al. [108] simulated this phenomenon to study the main forces that drive it. They concluded that the fundamental driver of high water saturations held near the fracture-matrix interface may be due to capillary end effects. These act near the interface between fracture and formation matrix to increase the water saturation beyond the saturation caused by imbibition. Elputranto et al. [108] therefore suggested that capillary end effect is a significant mechanism that must be considered when assessing potential of water-blocking effect in a formation. Future research may focus on experimental validation of this simulation work.

In order to effectively diagnose the predominant mechanism of face damage in fractures in tight sands, Li et al. [109] proposed a new experimental method. In their work, two mechanisms are suggested as mostly being responsible for fracture face damage; high capillary pressure and swelling of water sensitive clays. Li et al. [109] integrated pressure transmission and pressure decay methods to determine the predominant cause of fracture face damage. They concluded that their method is able to distinguish the cause of the key mechanism in fracture face damage. Though the method was effective in tight sand, it has yet to be tested on shales and other unconventional reservoir rock samples with high clay mineral content. Future research should focus on investigating the scope of application of this method for shale formations and other unconventional reservoir rocks with high clay compositions.

7.2 Fines migration

Fine particle migration is a major cause of fracture aperture blocking yet a very difficult phenomenon to study. Muggli et al. [110] introduced a simple, time saving experimental method to assess particle migration potential for different fracturing fluid compositions. They premised their method on the fact that behaviour of fine particles in the subsurface is a function of fracturing fluid composition. In their method, turbidity, capillary suction time behaviour and particle size distribution relative to the time and depth of particles are observed and recorded. Results from observations are then used to draw conclusions on the potential of particle migration and pore throat blocking. They tested their method using Eagle Ford Brine with and without additives. Low turbidity and capillary suction time were observed for brine without additives whiles higher values were recorded when additives were used. They explained that high turbidity was due to the inhibition of flocculation caused by the additives. Low capillary suction times therefore, may not always be desirable. In conclusion, they indicated that this experiment can be repeated using other fracturing fluid compositions to determine their impact on migration of fines.

7.3 Mineral dissolution and precipitation

Significant geochemical reactions are expected to occur during the shut-in period when fracturing fluid is in contact with formation. Wang et al. [111] studied rock-fluid interactions during the shut-in period to assess the water chemistry over the period using 15% HCL and water. They also assessed the possibility of scale formation in the reservoir based on these reactions. Their results showed ability of fracturing fluid to react with and dissolve formation minerals to increase permeability. However, these reactions also lead to release of ions into solution which may cause precipitation of scale-forming and permeability reduction minerals. Wang et al. [111] proposed their study as a way of understanding the long-term implications of rock-fluid interactions. Future research to expand the frontiers of this work should consider using different fracturing fluid compositions.

Furthermore, the impact of microstructural and geochemical interactions between fracturing fluids and fracture face in organic rich carbonates was studied by Liang et al. [112]. They concluded that 2% Potassium Chloride, though used to inhibit adverse reactions between fluids and clays minerals, may in fact increase the dissolution rates of carbonates thus increasing absolute permeability. This increase was observed to be more pronounced for slickwater relative to synthetic sea water. Their study is evidence that water-based fracturing fluids could be beneficial when used in some formation types such as organic-rich carbonates. Future research in this area should therefore be focused on understanding the mechanisms that cause faster dissolution in presence of Potassium Chloride.

7.4 Changes in mechanical properties

Juan et al. [113] investigated the relative impacts of slickwater and linear gel on mechanical properties of different rock types in the Permian basin. Reactions for these set of experiments were conducted at elevated temperature and pressure conditions, 190°F and 1000 psi respectively. Their findings indicated that: linear gel caused more mechanical (Young's Modulus) reduction, about 27% compared to slickwater, about 14%; Samples with higher contents of carbonates sustained more damage relative to low carbonate samples, with carbonate etching being the primary damage mechanism in slickwater whilst that in the linear gel is aggressive dissolution; Most carbonate dissolution happened within the first five hours of the reaction. This experiment provides critical information on reactions between rocks and fluids at elevated temperatures and could be repeated for other types of fracturing fluids in different formations to observe the responses.

Temperature differences between injected fracturing fluids and formation have been reported to exert significant mechanical impacts during fracturing. This observation has gained attention and has become focus area for researchers. Vena et al. [114] studied the impact of large temperature differentials between formation and invading fracturing fluids. They took particular interest in changes such as clay swelling, imbibition and other mechanisms that adversely affect formation permeability. Their results indicate an initially pronounced impact on stress regimes within the formation leading to development of micro-fractures which are sealed over time. Similar findings were obtained by Elputranto et al. [115] when they used high resolution simulation methods to assess the response of formation to fluid with high temperature and salinity differentials to formation. Since perpetual propagation or opening of these micro-fractures will greatly enhance permeability of a reservoir, future research should be focused on understanding the mechanisms that can sustain these micro-fractures.

Elputranto et al. [115] used high resolution simulated models to investigate the mechanical impact on the interface between hydraulic fracture and matrix due to reactions emanating from cold and low salinity fracturing fluid invading rock formation. They simulated the responses during the well shut-in period and flowback and production periods. Their results show that thermo-elastic effects are generated in the formation that lead to increased permeability which is short lived. Based on results from this work, future research will focus on how to sustain and possibly allow better propagation of these short-lived fractures created due to thermo-elastic effects of fracturing fluid interaction with formation. Achieving this will lead to significantly improved permeability and production.

7.5 Alternative fracturing fluids

Li et al. [116] conducted an experiment on the use of CO_2 as pre-fracturing fluid during hydraulic fracturing in tight gas formations. They aimed to confirm that the combination of CO_2 and water during hydraulic fracturing operations could help harness benefits of both fluids especially at locations with low water availability. This research was undertaken for subsequent application in the Loess plateau of Ordos basin in China. Results from experiments showed improved permeability due to dissolution of carbonates and clay minerals by CO_2. They also found that CO_2 interacts with hydrocarbons and provides additional impacts in terms of improving hydrocarbon properties to enhance relative permeability, thus providing increased productivity. More experiments should be conducted to ascertain the optimum use of CO_2 and water combinations for fracturing.

Adverse environmental footprints of hydraulic fracturing operations have also necessitated research to find innovative ways of mitigating these impacts. Ellafi et al. [117] investigated the possibilities of re-using produced water as fracturing fluid. They justified their research by drawing attention to the large volumes of fresh water used in hydraulic fracturing operations. Some helpful statistics quoted to buttress their points include the following: Texans waste about 2% of water demand on fracturing jobs [118]; the amount of water withdrawn from the Missouri River for hydraulic fracturing in 2018 alone, was about 1.269×10^{10} gals, an estimated 10.1% of North Dakota water consumption [119].

Acknowledgements

The authors would like to acknowledge this study was made possible by The US DOE Award DE-FE0031776 from the Office of Fossil Energy. We also thank Ben Chapman of College of Engineering, Architecture and Technology (CEAT), OSU for helping with the graphics. We are grateful to our Team Members in Hydraulic Barrier and Geomimicry Materials at OSU: Allan Katende, Cody Massion, Chris Grider and Vamsi Vissa for their support.

Author details

Gabriel Adua Awejori and Mileva Radonjic*
Oklahoma State University (OSU), Stillwater, USA

*Address all correspondence to: mileva.radonjic@okstate.edu

IntechOpen

References

[1] Drits VA, Lindgreen H, Sakharov BA, Jakobsen HJ, Salyn AL, Dainyak LG. Tobelitization of smectite during oil generation in oil-source shales. Application to North Sea illite-tobelite-smectite-vermiculite. Clays Clay Miner. 2002;50(1):82-98.

[2] Weaver CE. The effects and geologic significance of potassium "fixation" by expandable clay minerals derived from muscovite, biotite, chlorite, and volcanic material. Am Mineral. 1958;

[3] Weaver CE. Possible Uses of Clay Minerals in the Search for Oil. In: Clays and Clay Minerals. 1960. p. 214-27.

[4] Sarkissian SG. Mineralogic Composition of Clays in Petroliferous Deposits of the USSR: Some Data on Geology and Mineralogy and Utilization of Clays in the USSR. Reports on International Meeting on Clays in Brussels. 1958.

[5] Hamilton PJ. K-Ar Dating of Illite in Hydrocarbon Reservoirs. Clay Miner. 1989;

[6] Jiang S. Clay Minerals from the Perspective of Oil and Gas Exploration. In: Clay Minerals in Nature - Their Characterization, Modification and Application. 2012. p. 21-38.

[7] Kelly J, Parnell J, Chen HH. Application of fluid inclusions to studies of fractured sandstone reservoirs. In: Journal of Geochemical Exploration. 2000. p. 705-9.

[8] Liewig N, Clauer N, Sommer F. Rb-Sr AND K-Ar Dating Of Clay Diagenesis In Jurassic Sandstone Oil Reservoir, North Sea. Am Assoc Pet Geol Bull. 1987;71(12):1467-74.

[9] Yariv S. Organophilic Pores as Proposed Primary Migration Media for Hydrocarbons In Argillaceous Rocks. Clay Sci. 1976;

[10] Lal M. Shale stability: Drilling fluid interaction and shale strength. In: Society of Petroleum Engineers - SPE Asia Pacific Oil and Gas Conference and Exhibition 1999, APOGCE 1999. 1999.

[11] Durand C, Forsans T, Ruffet C, Onaisi A, Audibert A. Influence of clays on borehole stability: a literature survey. Part one: occurrence of drilling problems, physico-chemical description of clays and of their interaction with fluids. Rev - Inst Fr du Pet. 1995;

[12] Durand C, Forsans T, Ruffet C, Onaisi A, Audibert A. Influence of clays on borehole stability: a literature survey part two: mechanical description and modelling of clays and shales drilling practices versus laboratory simulations. Rev - Inst Fr du Pet. 1995;50(3):353-69.

[13] Rahman MK, Suarez YA, Chen Z, Rahman SS. Unsuccessful hydraulic fracturing cases in Australia: Investigation into causes of failures and their remedies. J Pet Sci Eng. 2007;57(1-2):70-81.

[14] van Oort E. Physico-chemical stabilization of shales. In: Proceedings - SPE International Symposium on Oilfield Chemistry. 1997. p. 523-38.

[15] Zhou Z, Gunter WD, Kadatz B, Cameron S. Effect of clay swelling on reservoir quality. J Can Pet Technol. 1996;35(7):18-23.

[16] Zhou ZJ, Gunter WD, Jonasson RG. Controlling formation damage using clay stabilizers: A review. In: Annual Technical Meeting 1995, ATM 1995. 1995.

[17] Shukla R, Ranjith PG, Choi SK, Haque A, Yellishetty M, Hong L. Mechanical behaviour of reservoir rock

under brine saturation. Rock Mech Rock Eng. 2013;46(1):83-93.

[18] Alalli A, Li Q, Jew A, Kohli A, Bargar J, Zoback M, et al. Effects of hydraulic fracturing fluid chemistry on shale matrix permeability. In: SPE/AAPG/SEG Unconventional Resources Technology Conference 2018, URTC 2018. 2018.

[19] Buller D, Hughes S, Market J, Petre E, Spain D, Odumosu T. Petrophysical evaluation for enhancing hydraulic stimulation in horizontal shale gas wells. In: Proceedings - SPE Annual Technical Conference and Exhibition. 2010. p. 431-51.

[20] Radonjic M, Luo G, Wang Y, Achang M, Cains J, Katende A, et al. Integrated Microstructural Characterisation of Caney Shale, OK. 2020;1-18.

[21] Olabode A, Radonjic M. Characterization of shale cap-rock nano-pores in geologic CO2 containment. Environ Eng Geosci. 2014;20(4):361-70.

[22] Olabode A, Radonjic M. Fracture Conductivity Modelling in Experimental Shale Rock Interactions with Aqueous CO2. Energy Procedia [Internet]. 2017;114(November 2016):4494-507. Available from: http://dx.doi.org/10.1016/j.egypro.2017.03.1610

[23] Busch A, Alles S, Gensterblum Y, Prinz D, Dewhurst DN, Raven MD, et al. Carbon dioxide storage potential of shales. Int J Greenh Gas Control. 2008;2(3):297-308.

[24] Busch A, Amann A, Bertier P, Waschbusch M, Krooss BM. The significance of caprock sealing integrity for CO2 storage. In: Society of Petroleum Engineers - SPE International Conference on CO2 Capture, Storage, and Utilization 2010. 2010. p. 300-7.

[25] Olabode A, Radonjic M. Experimental investigations of caprock integrity in CO2 sequestration. In: Energy Procedia. 2013. p. 5014-25.

[26] Olabode A, Radonjic M. Shale Caprock/Acidic Brine Interaction in Underground CO2 Storage. J Energy Resour Technol. 2014;136(4):1-6.

[27] Olabode A, Radonjic M. Diagenetic influence on fracture conductivity in tight shale and CO2 sequestration. Energy Procedia. 2014;63:5021-31.

[28] Jeon PR, Choi J, Yun TS, Lee CH. Sorption equilibrium and kinetics of CO2 on clay minerals from subcritical to supercritical conditions: CO2 sequestration at nanoscale interfaces. Chem Eng J. 2014;255:705-15.

[29] Espinoza DN, Santamarina JC. Clay interaction with liquid and supercritical CO 2: The relevance of electrical and capillary forces. Int J Greenh Gas Control. 2012;10:351-62.

[30] Du H, Carpenter K, Hui D, Radonjic M. Microstructure and micromechanics of shale rocks: Case study of marcellus shale. Facta Univ Ser Mech Eng. 2017;15(2):331-40.

[31] Aksu I, Bazilevskaya E, Karpyn ZT. Swelling of clay minerals in unconsolidated porous media and its impact on permeability. GeoResJ. 2015;

[32] Hamdi N, Srasra E. Acid-base properties of organosmectite in aqueous suspension. Appl Clay Sci. 2014;

[33] Wang LL, Zhang GQ, Hallais S, Tanguy A, Yang DS. Swelling of Shales: A Multiscale Experimental Investigation. Energy and Fuels. 2017;

[34] Hughes R V. The application of modern clay concepts to oilfield development. In: Drilling and Production Practice 1950. 1950.

[35] Bergaya F, Lagaly G, Vayer M. Cation and Anion Exchange. In: Developments in Clay Science. 2013. p. 333-59.

[36] Bush DC, Jenkins RE. CEC determinations by correlations with adsorbed water. In: SPWLA 18th Annual Logging Symposium 1977. 1977.

[37] Davidson DT, Sheeler JB. Cation Exchange Capacity of Loess and its Relation to Engineering Properties. In: Symposium on Exchange Phenomena in Soils. 2009. p. 10-10-9.

[38] Hill HJ, Milburn JD. Effect of clay and water salinity on electrochemical behavior of reservoir rocks. SPE Repr Ser. 2003;(55):31-8.

[39] Thomas EC. Determination of Qv From Membrane Potential Measurements on Shaly Sands. JPT, J Pet Technol. 1976;28:1087-96.

[40] Worthington AE. an Automated Method for the Measurement of Cation Exchange Capacity of Rocks. Geophysics. 1973;38(1):140-53.

[41] Cheng K, Heidari Z. A new method for quantifying cation exchange capacity in clay minerals. Appl Clay Sci. 2018;161:444-55.

[42] Cheng K, Heidari Z. A new method for quantifying cation exchange capacity in clay minerals. In: SPWLA 58th Annual Logging Symposium 2017. 2017.

[43] Botan A, Rotenberg B, Marry V, Turq P, Noetinger B. Carbon dioxide in montmorillonite clay hydrates: Thermodynamics, structure, and transport from molecular simulation. J Phys Chem C. 2010;114(35):14962-9.

[44] Doostmohammadi R, Moosavi M. Swelling of weak rocks, effective parameters and controlling methods.

In: ISRM International Symposium - 5th Asian Rock Mechanics Symposium 2008, ARMS 2008. 2008. p. 247-53.

[45] Salles F, Beurroies I, Bildstein O, Jullien M, Raynal J, Denoyel R, et al. A calorimetric study of mesoscopic swelling and hydration sequence in solid Na-montmorillonite. Appl Clay Sci. 2008;39(3-4):186-201.

[46] Warr L, Berger J. Hydration of bentonite in natural waters: Application of "confined volume" wet-cell X-ray diffractometry. Phys Chem Earth. 2007;32(1-7):247-58.

[47] Patel A, Stamatakis E, Young S, Friedheim J. Advances in inhibitive water-based drilling fluids - Can they replace oil-based muds? In: Proceedings - SPE International Symposium on Oilfield Chemistry. 2007. p. 614-21.

[48] Mayerhofer MJ, Lolon EP, Rightmire C, Walser D, Cipolla CL, Warplnskl NR. What is stimulated reservoir volume? SPE Prod Oper. 2010;

[49] Yuan B, Su Y, Moghanloo RG, Rui Z, Wang W, Shang Y. A new analytical multi-linear solution for gas flow toward fractured horizontal wells with different fracture intensity. J Nat Gas Sci Eng. 2015;

[50] Rickards AR, Brannon HD, Wood WD, Stephenson CJ. High strength, ultralightweight proppant lends new dimensions to hydraulic fracturing applications. SPE Prod Oper. 2006;

[51] Montgomery C. Fracturing fluids. In: ISRM International Conference for Effective and Sustainable Hydraulic Fracturing 2013. 2013.

[52] Ribeiro LH, Sharma MM. Fluid selection for energized fracture treatments. In: Society of Petroleum

Engineers - SPE Hydraulic Fracturing Technology Conference 2013. 2013.

[53] Agency USEP. Proceedings of the Technical Workshops for the Hydraulic Fracturing Study : Fate and Transport. Epa 600. 2011;

[54] Cikes M, Cubric S, Moylashov MR. Formation damage prevention by using an oil-based fracturing fluid in partially depleted oil reservoirs of Western Siberia. In: Proceedings - SPE International Symposium on Formation Damage Control. 1998.

[55] Perfetto R, Melo RCB, Martocchia F, Lorefice R, Ceccarelli R, Tealdi L, et al. Oil-based fracturing fluid: First results in West Africa onshore. In: Society of Petroleum Engineers - International Petroleum Technology Conference 2013, IPTC 2013: Challenging Technology and Economic Limits to Meet the Global Energy Demand. 2013.

[56] Hlidek BT, Meyer RK, Yule K, Wittenberg J. A case for oil-based fracturing fluids in Canadian Montney unconventional gas development. In: Proceedings - SPE Annual Technical Conference and Exhibition. 2012.

[57] Gupta DVS, Leshchyshyn TT. CO2 energized hydrocarbon fracturing fluid: History & field application in tight gas wells in the rock creek gas formation. In: SPE Latin American and Caribbean Petroleum Engineering Conference Proceedings. 2005.

[58] Vezza M, Martin M, Thompson JE, DeVine C. Morrow Production Enhanced by New, Foamed, Oil-Based Gel Fracturing Fluid Technology. In: Proceedings - SPE Production Operations Symposium. 2001.

[59] Freeman ER, Abel JC, Chin Man Kim, Heinrich C. Stimulation Technique Using Only Nitrogen. JPT, J Pet Technol. 1983;

[60] Rogala A, Krzysiek J, Bernaciak M, Hupka J. Non-aqueous fracturing technologies for shale gas recovery. Physicochem Probl Miner Process. 2013;49(1):313-21.

[61] Gu M, Dao E, Mohanty KK. Investigation of ultra-light weight proppant application in shale fracturing. Fuel. 2015;

[62] Gandossi L. An overview of hydraulic fracturing and other formation stimulation technologies for shale gas production. JRC Tech Reports. 2013;

[63] Ishida T, Nagaya Y, Inui S, Aoyagi K, Nara Y, Chen Y, et al. AE monitoring of hydraulic fracturing experiments conducted using CO2 and water. In: ISRM International Symposium - EUROCK 2013. 2013.

[64] Middleton RS, Carey JW, Currier RP, Hyman JD, Kang Q, Karra S, et al. Shale gas and non-aqueous fracturing fluids: Opportunities and challenges for supercritical CO2. Appl Energy. 2015;

[65] Handy LL. Determination of Effective Capillary Pressures for Porous Media from Imbibition Data. Trans AIME. 1960;

[66] Bear J. Dynamics of Fluids in Porous Media. Soil Sci. 1975;

[67] Bennion DB, Thomas FB, Imer D, Ma T. Low permeability gas reservoirs and formation damage - tricks and traps. In: SPE Proceedings - Gas Technology Symposium. 2000.

[68] Hoffman ME. Reservoirs that are not in capillary pressure equilibrium. In: Unconventional Resources Technology Conference 2013, URTC 2013. 2013.

[69] Dutta R. Laboratory study of fracturing fluid migration due to spontaneous imbibition in fractured

tight formations. In: Proceedings - SPE Annual Technical Conference and Exhibition. 2012.

[70] Bertoncello A, Wallace J, Blyton C, Honarpour M, Kabir CS. Imbibition and water blockage in unconventional reservoirs: Well-management implications during flowback and early production. SPE Reserv Eval Eng. 2014;

[71] Bostrom N, Chertov M, Pagels M, Willberg D, Chertova A, Davis M, et al. The time-dependent permeability damage caused by fracture fluid. In: SPE - European Formation Damage Conference, Proceedings, EFDC. 2014.

[72] Ma S, Zhang X, Morrow NR. Influence of fluid viscosity on mass transfer between rock matrix and fractures. J Can Pet Technol. 1999;

[73] Blair PM. Calculation of Oil Displacement by Countercurrent Water Imbibition. Soc Pet Eng J. 1964;

[74] Li K, Chow K, Horne RN. Effect of Initial Water Saturation on Spontaneous Water Imbibition. In: SPE Western Regional/AAPG Pacific Section Joint Meeting. 2002.

[75] Cil M, Reis JC, Miller MA, Misra D. An examination of countercurrent capillary imbibition recovery from single matrix blocks and recovery predictions by analytical matrix/fracture transfer functions. In: Proceedings - SPE Annual Technical Conference and Exhibition. 1998.

[76] Zhou X, Morrow NR, Ma S. Interrelationship of wettability, initial water saturation, aging time, and oil recovery by spontaneous imbibition and waterflooding. SPE J. 2000;

[77] Viksund B, Morrow N, Ma S. Initial water saturation and oil recovery from chalk and sandstone by spontaneous imbibition. ... Symp Soc 1998;

[78] Akin S, Schembre JM, Bhat SK, Kovscek AR. Spontaneous imbibition characteristics of diatomite. J Pet Sci Eng. 2000;

[79] Pooladi-Darvish M, Firoozabadi A. Experiments and modelling of water injection in water-wet fractured porous media. In: Annual Technical Meeting 1998, ATM 1998. 1998.

[80] Zhou Z, Abass H, Li X, Bearinger D, Frank W. Mechanisms of imbibition during hydraulic fracturing in shale formations. J Pet Sci Eng. 2016;

[81] Zhou Z, Li X, Teklu TW. A Critical Review of Osmosis-Associated Imbibition in Unconventional Formations. Energies. 2021;

[82] Bazin B, Bekri S, Vizika O, Herzhaft B, Aubry E. Fracturing in tight gas reservoirs: Application of special-core-analysis methods to investigate formation-damage mechanisms. SPE J. 2010;

[83] Chakraborty N, Karpyn ZT, Liu S, Yoon H. Permeability evolution of shale during spontaneous imbibition. J Nat Gas Sci Eng. 2017;

[84] Zhang D, Kang Y, Selvadurai APS, You L, Tian J. The role of phase trapping on permeability reduction in an ultra-deep tight sandstone gas reservoirs. J Pet Sci Eng. 2019;

[85] Shanley KW, Cluff RM, Robinson JW. Factors controlling prolific gas production from low-permeability sandstone reservoirs: Implications for resource assessment, prospect development, and risk analysis. Am Assoc Pet Geol Bull. 2004;

[86] Kamath J, Laroche C. Laboratory-based evaluation of gas well deliverability loss caused by water blocking. SPE J. 2003;

[87] Ellis BR, Fitts JP, Bromhal GS, McIntyre DL, Tappero R, Peters CA. Dissolution-driven permeability reduction of a fractured carbonate caprock. Environ Eng Sci. 2013;

[88] Nogues JP, Fitts JP, Celia MA, Peters CA. Permeability evolution due to dissolution and precipitation of carbonates using reactive transport modeling in pore networks. Water Resour Res. 2013;

[89] Ferrage E, Lanson B, Sakharov BA, Drits VA. Investigation of smectite hydration properties by modeling experimental X-ray diffraction patterns: Part I: Montmorillonite hydration properties. Am Mineral. 2005;

[90] Likos WJ, Lu N. Pore-scale analysis of bulk volume change from crystalline interlayer swelling in Na+− and Ca2+− smectite. Clays Clay Miner. 2006;

[91] Gonçalvès J, Rousseau-Gueutin P, De Marsily G, Cosenza P, Violette S. What is the significance of pore pressure in a saturated shale layer? Water Resour Res. 2010;

[92] Saiyouri N, Hicher PY, Tessier D. Microstructural approach and transfer water modelling in highly compacted unsaturated swelling clays. Mech Cohesive-Frictional Mater. 2000;

[93] Laird DA. Influence of layer charge on swelling of smectites. Appl Clay Sci. 2006;

[94] Santos H, Diek A, Da Fontoura S, Roegiers JC. Shale reactivity test: a novel approach to evaluate shale-fluid interaction. Int J rock Mech Min Sci Geomech Abstr. 1997;

[95] Gupta A, Xu M, Dehghanpour H, Bearinger D. Experimental investigation for microscale stimulation of shales by water imbibition during the shut-in periods. In: Society of Petroleum Engineers - SPE Unconventional Resources Conference 2017. 2017.

[96] Dehghanpour H, Lan Q, Saeed Y, Fei H, Qi Z. Spontaneous imbibition of brine and oil in gas shales: Effect of water adsorption and resulting microfractures. Energy and Fuels. 2013;

[97] Chenevert ME. Shale Alteration by Water Adsorption. JPT, J Pet Technol. 1970;

[98] Du H, Radonjic M, Chen Y. Microstructure and micro-geomechanics evaluation of Pottsville and Marcellus shales. J Pet Sci Eng. 2020;

[99] Akrad O, Miskimins J, Prasad M. The effects of fracturing fluids on shale rock mechanical properties and proppant embedment. In: Proceedings - SPE Annual Technical Conference and Exhibition. 2011.

[100] Corapcioglu H, Miskimins JL, Prasad M. Fracturing fluid effects on young's modulus and embedment in the Niobrara formation. In: Proceedings - SPE Annual Technical Conference and Exhibition. 2014.

[101] LaFollette RF, Carman PS. Proppant diagenesis: Results so far. In: SPE Unconventional Gas Conference 2010. 2010.

[102] Carman PS, Lant KS. Making the case for shale clay stabilization. In: SPE Eastern Regional Meeting. 2010.

[103] You L, Xie B, Yang J, Kang Y, Han H, Wang L, Yang B. Mechanism of fracture damage induced by fracturing fluid flowback in shale gas reservoirs. Natural Gas Industry B 6 366-373. 2019

[104] Chapman EC, Capo RC, Stewart BW, Kirby CS, Hammack RW, Schroeder KT, et al. Geochemical and strontium isotope characterization of produced waters from marcellus shale natural gas extraction. Environ Sci Technol. 2012;

[105] Phan TT, Capo RC, Stewart BW, Graney JR, Johnson JD, Sharma S, et al. Trace metal distribution and mobility in drill cuttings and produced waters from Marcellus Shale gas extraction: Uranium, arsenic, barium. Appl Geochemistry. 2015;

[106] Marcon V, Joseph C, Carter KE, Hedges SW, Lopano CL, Guthrie GD, et al. Experimental insights into geochemical changes in hydraulically fractured Marcellus Shale. Appl Geochemistry. 2017;

[107] Wilke FDH, Vieth-Hillebrand A, Naumann R, Erzinger J, Horsfield B. Induced mobility of inorganic and organic solutes from black shales using water extraction: Implications for shale gas exploitation. Appl Geochemistry. 2015;

[108] Elputranto R, Akkutlu IY. Near fracture capillary end effect on shale gas and water production. In: SPE/AAPG/SEG Unconventional Resources Technology Conference 2018, URTC 2018. 2018.

[109] Li H, Li B, Zhou F, Zhang D, Zhang Y, Xian B, et al. A new experimental approach for hydraulic fracturing fluid optimization: Especially focus on the ultra-deep tight gas formation. In: 53rd US Rock Mechanics/Geomechanics Symposium. 2019.

[110] de Araujo Muggli I, Chellappah K, Collins IR. An Experimental Approach to Assess the Dispersion of Shale in Fracturing Fluids. SPE Prod Oper. 2020;

[111] Wang W, Wei W, Leach D, Yan C, Spilker K. Rock-Fluid Interaction and Its Applications in Unconventional Production. In 2020.

[112] Liang F, Zhang J, Liu HH, Bartko KM. Multiscale experimental studies on interactions between aqueous-based fracturing fluids and tight organic-rich carbonate source rocks. In: SPE Reservoir Evaluation and Engineering. 2019.

[113] Acosta JC, Dang S, Curtis M, Sondergeld C, Rai C. Fracturing Fluids Effect on Mechanical Properties in Shales. In 2020.

[114] Eveline VF, Santos LP, Yucel Akkutlu I. Thermally-induced secondary fracture development in shale formations during hydraulic fracture water invasion and clay swelling. In: Society of Petroleum Engineers - SPE Europec Featured at 81st EAGE Conference and Exhibition 2019. 2019.

[115] Elputranto R, Cirdi AP, Yucel Akkutlu I. Formation Damage Mechanisms Due to Hydraulic Fracturing of Shale GasWells. In: Society of Petroleum Engineers - SPE Europec Featured at 82nd EAGE Conference and Exhibition. 2020.

[116] Li L, Su Y, Chen Z, Fan L, Tang M, Tu J. Experimental investigation on EOR and flowback rate of using supercritical CO2 as pre-fracturing fluid in tight oil reservoir. In: Society of Petroleum Engineers - SPE Asia Pacific Oil and Gas Conference and Exhibition 2020, APOG 2020. 2020.

[117] Ellafi A, Jabbari H, Tomomewo OS, Mann MD, Geri MB, Tang C. Future of hydraulic fracturing application in terms of water management and environmental issues: A critical review. In: Society of Petroleum Engineers - SPE Canada Unconventional Resources Conference 2020, URCC 2020. 2020.

[118] Environment Texas Research and Policy Center. Keeping water in our rivers; strategies for conserving limited water supplies. http://environmenttexas.org/reports/txe/keeping-water-our-rivers (accessed 5.18.21). 2013

[119] NDSWC. North Dakota Fracking and Water use Facts. http://www.http://www.swc.nd.gov/pdfs/fracking_water_use.pdf (accessed 5.18.21). 2019

Section 2

The Gas Flow Model

Chapter 6

Mechanism, Model, and Upscaling of the Gas Flow in Shale Matrix: Revisit

Zhiming Hu, Yaxiong Li and Yanran Li

Abstract

Shale gas accounts for an increasing proportion in the world's oil and gas supply, with the properties of low carbon, clean production, and huge potential for the compensation for the gradually depleted conventional resources. Due to the ubiquitous nanopores in shale matrix, the nanoscale gas flow becomes one of the most vital themes that are directly related to the formulation of shale gas development schemes, including the optimization of hydraulic fracturing, horizontal well spacing, etc. With regard to the gas flow in shale matrix, no commonly accepted consensus has been reached about the flow mechanisms to be considered, the coupled flow model in nanopores, and the upscaling method for its macroscopic form. In this chapter, the propositions of wall-associated diffusion, a physically sound flow mechanism scheme, a new coupled flow model in nanopores, the upscaling form of the proposed model, and the translation of lab-scale results into field-scale ones aim to solve the aforementioned issues. It is expected that this work will contribute to a deeper understanding of the intrinsic relationship among various flow mechanisms and the extension of the flow model to full flow regimes and to upscaling shale matrix, thus establishing a unified model for better guiding shale gas development.

Keywords: shale gas, diffusion, viscous flow, coupling coefficient, generalized model, pore size distribution, macroscopic form

1. Introduction

Shale gas refers to a kind of self-generating and self-preserving natural gas, which gathers mainly in a free or adsorbed state in the organic-rich dark shale or high-carbon mud shale [1]. With vast reserves and the potential to offset the gradually depleted conventional resources worldwide and cut down carbon emissions at the same time, shale gas is playing an increasingly important role in ensuring global energy safety. Because shale matrix is characterized by various nanopores, where the gas flow is of high nonlinearity and complexity, an in-depth study of the mathematical model for the gas flow capacity in shale matrix is in urgent demand.

The mechanisms considered in different literature are listed in **Table 1**. It is obvious that opinions vary greatly on the flow mechanism scheme applied. The noteworthy aspects include the following: what the relationship among the various

Literature	Mechanisms considered
Klinkenberg [2]	Slip flow
Javadpour [3], Haghshenas et al. [4], Wu et al. [5], Sun et al. [6]	Knudsen diffusion and slippage
Veltzke and Thöming [7]	Viscous flow and Knudsen diffusion
Li et al. [8]	Continuum flow, slip flow, transition flow, and free molecular flow
Mi et al. [9]	Diffusion and slippage, where the form of diffusion varies according to the Knudsen number range, including Fick diffusion, transitional diffusion, and Knudsen diffusion
Song et al. [10]	Viscous flow, Knudsen diffusion, and surface diffusion, with surface diffusion not considered for inorganic pores

Table 1.
Different flow mechanism schemes in literature.

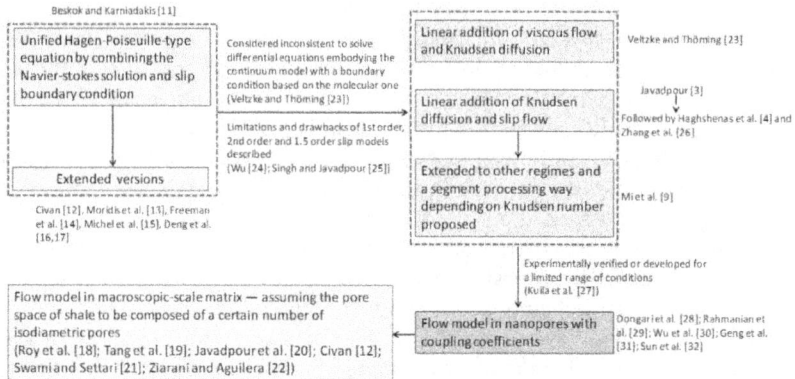

Figure 1.
A brief summary of the common methodology used in different research [11–32].

flow mechanisms of shale gas, e.g., slippage, Fick diffusion, Knudsen diffusion, etc., is; whether there is a repeated superposition of these mechanisms for specific flow calculation; and how to deal with the relationship among the various flow mechanisms, etc. There is no clear answer to these problems in current literature.

Figure 1 shows the common research methodology of the flow models used in different literature. It indicates that because the method of the continuum model with a boundary condition based on the molecular one is considered inconsistent and the limitations and drawbacks of first-order, second-order, and 1.5-order slip models are described, some studies, which are listed in **Figure 1**, are inclined to add related flow mechanisms linearly. Furthermore, the mathematical models of viscous flow and various types of diffusion do not fully agree with common flow cognition as these theories and models were experimentally verified or developed for a limited range of conditions [27]. For this reason, coupling coefficients are introduced to rectify this kind of limitation, so as to enhance the correspondence between the flow model and Knudsen number (Kn). Finally, because the secondhand average method, e.g., assuming the pore space of shale to be composed of a certain number of isodiametric pores regardless of the pore size distribution, is widely used in the research of shale gas flow, more explicit means, like taking the existence of various pore sizes in shale into account, should be adopted for transforming the flow model in nanopores to that in macroscopic-scale shale matrix.

Based on the literature survey for shale gas flow in shale matrix, we know that the flow mechanism scheme with its corresponding coupling method is very crucial and has not yet been solved. In addition, although the integration method using specific functions has been proposed to facilitate the consideration of various pore sizes in shale matrix, real shale experiments are rarely involved to realize this point with definitely determined parameters.

Firstly, in this chapter, the concept of wall-associated diffusion is presented to clarify the relationship between slippage effect and several types of diffusion. Secondly, a physically sound flow mechanism scheme, which considers both division of mechanical mechanisms in nanopores and partition of flow space, has been proposed by virtue of the proposition of wall-associated diffusion. Thirdly, the coupling coefficients corresponding to the flow mechanisms considered are deduced to comply with the basic flow regime cognition, so as to establish a new coupled flow model in nanopores. Fourthly, the pore size distribution experiments for real shale samples from a gas field are utilized to realize the upscaling transformation of the flow model in nanopores into that in the macroscopic-scale shale matrix, with definitely determined fitting parameters for the establishment of a unified model for the gas flow prediction in shale matrix. Finally, a case study is presented to show how the lab-scale results are translated into field-scale ones.

2. Flow mechanisms in gas-shale matrix

There are many types of flow mechanisms in shale matrix, including slippage effect, Fick diffusion, transition diffusion, Knudsen diffusion, surface diffusion, etc. It can be seen from the literature survey in Section 1 that different flow mechanism schemes have formed aiming at establishing a calculation model to properly characterize the nanoscale shale gas flow. There may be views that the more flow mechanisms are taken into account, the more precise the established models are. However, this is not the opinion in this chapter.

As is known, Klinkenberg [33] first discovered in 1941 the phenomenon that, when measuring the gas permeability of rock, not only the measurement result is higher than the liquid measurement value but also it has strong pressure dependence and attributed it to the slippage behavior of gas in the rock pores. Specifically, gas slippage refers to the phenomenon that the near-wall gas molecules move relative to the wall surface when flowing through the medium channels [34]. In essence, the gas slip flow results from the interaction of gas molecules and pore walls, so the gas molecules in the vicinity of walls are in motion and contribute an additional flux, which is macroscopically characterized by the non-zero gas velocities on channel walls, thus resulting in slip flow [35, 36]. The jump model assumes that the adsorbed gas molecules jump from one adsorption site to the adjacent adsorption site on the pore surface, which is considered to be suitable for the research on the surface diffusion of the adsorbed gas in shale nanopores [37]. Meanwhile, when the molecular mean free path is obviously larger than the pore diameter, the gas-wall collision dominates, and the collision between gas molecules is secondary, which is characterized by Knudsen diffusion [9, 38, 39].

In brief, both Knudsen diffusion and surface diffusion lead to non-zero moving speeds of the gas molecules around walls. Furthermore, from the viewpoint of microscopic motion mechanisms, they are both related to gas–solid interactions, which is consistent with slippage phenomenon in essence. Therefore, a new concept named "wall-associated diffusion" [40] is proposed, which characterizes the overall role of surface diffusion and Knudsen diffusion, as shown in **Figure 2**.

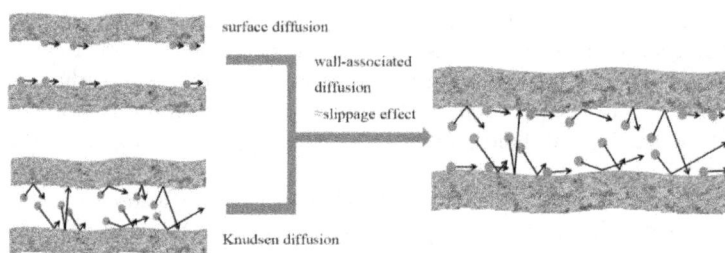

Figure 2.
Relationship between wall-associated diffusion and slippage effect [40].

The proposition of wall-associated diffusion has practical significance and multiple research significance as follows [40].

To begin with, in terms of mechanical mechanisms, since wall-associated diffusion describes the diffusion mechanisms of shale gas related to gas-wall interactions, it bridges the relationship between slippage effect and several types of diffusion, which prevents reduplicated superposition of shale gas flow mechanisms in nanoscale pores. This is where the practical significance lies. Besides, wall-associated diffusion can be regarded as a detailed form of slippage effect, dividing slippage effect into two distinct parts, i.e., surface diffusion and Knudsen diffusion. The two parts differ obviously in their mechanical mechanisms and motion patterns. Accordingly, the research significance of wall-associated diffusion involves not only the function of morphological descriptions but also the possibility of slip phenomenon research by different mechanical mechanisms. Lastly, another research significance is that wall-associated diffusion breaks through the limitation that the concept of slippage does not apply for high Knudsen number, with, however, the fact that wall effects still contribute to gas flow for high Knudsen number. Therefore, in extremely small nanopores, for example, where slip flow regime is not applicable, the wall-associated diffusion derived from physical morphology can well be used to explore the so-called slip phenomenon in other flow regimes apart from slip flow regime.

By virtue of the concept of wall-associated diffusion, the flow mechanism scheme used in this work is to be discussed next.

There is no doubt that all the mechanisms, such as continuum flow, slip flow, Knudsen diffusion, bulk diffusion, etc., have been studied in previous literature for the exploration of shale gas flow. However, it is a determinative flow mechanism scheme that is vital. According to the literature survey, apart from combining the Navier-Stokes solution with slip boundary condition whose deficiency has been mentioned in Section 1, there is also a trend in literature to assume a combination of certain flow mechanisms and check the consistency of the model results with experimental data. This method is favorable from an engineering point of view but meanwhile leads to the status that coincidence often exists and no commonly accepted consensus has formed currently. In this work, we discuss the issue physically. Firstly, due to the multiple advantages of wall-associated diffusion over the concept of slippage effect, slippage effect is replaced with wall-associated diffusion in the following discussion. On the one hand, the flow space in nanopores can be divided into two parts: the bulk phase region and the Knudsen layer [41]. On the other hand, the microscopic mechanical mechanisms can be divided into the gas–gas and gas-wall interactions. If a new comprehensive flow scheme, including viscous flow and bulk diffusion which belong to bulk phase flow and surface diffusion and Knudsen diffusion which are associated with gas-wall interactions causing non-zero flow velocities near pore walls, is proposed, the considerations of the division of flow space and mechanical mechanisms can be both realized.

It should be noted that with the help of the methodology applied here, some flow mechanisms that are easily omitted are now included, such as bulk diffusion, an important diffusion process which is controlled by a mechanical mechanism obviously different from Knudsen diffusion. Furthermore, because the individual flow expressions, e.g., those for viscous flow and diffusion, were experimentally verified or developed for a limited range of conditions [27], the proposed physical flow mechanism scheme avoids unnecessary attempts to fit the mathematical models to experimental data so as to determine which flow mechanisms should be considered, laying a solid foundation for the research on the coupled flow model in nanopores discussed below.

To conclude, taking both division of mechanical mechanisms in nanopores and partition of flow space into account, viscous flow and bulk diffusion, which belong to bulk phase flow and result from gas–gas interactions, and surface diffusion and Knudsen diffusion, which are associated with gas-solid interactions and result in non-zero flow velocities near pore walls, are included in the proposed flow mechanism scheme.

3. Coupled model of shale gas flow in nanopores

Based on the flow scheme proposed in Section 2, the flow mechanisms considered include viscous flow, bulk diffusion, surface diffusion, and Knudsen diffusion. Considering the influence of adsorption layers, in which the system is assumed to reach dynamic adsorption equilibrium state instantaneously, the mass flow of the four mechanisms can be expressed, respectively, as:

$$N_D = -\frac{10^{-36}\pi\rho_{avg}}{8\mu}\left(r_{in} - \frac{pd_m}{p_L+p}\right)^4 \frac{dp}{dl} \tag{1}$$

$$N_b = N_F = -\frac{10^{-9}Mk_B}{3R\mu d_m}\left(r_{in} - \frac{pd_m}{p_L+p}\right)^2 \frac{dp}{dl} \tag{2}$$

$$N_K = -\frac{2\times10^{-27}}{3}\left(\frac{8\pi M}{RT}\right)^{0.5}\left(r_{in} - \frac{pd_m}{p_L+p}\right)^3 \frac{dp}{dl} \tag{3}$$

$$N_s = -0.016\times10^{-22}\times\exp\left(-\frac{0.45q}{RT}\right)\frac{\rho_s M}{pV_{std}}\frac{q_L p}{p_L+p}\cdot\frac{1-\phi_{co}}{\phi_{co}}\pi r_{in}^2\frac{dp}{dl} \tag{4}$$

where N_D = viscous mass flow in a pipe, kg·s^{-1}.
N_b = mass flow of bulk diffusion, kg·s^{-1}.
N_F = mass flow of Fick diffusion, kg·s^{-1}.
N_K = mass flow of Knudsen diffusion, kg·s^{-1}.
N_s = mass flow of surface diffusion, kg·s^{-1}.
r_{in} = inner radius of a pipe, nm.
ρ_{avg} = density of gas at average pressure of inlet and outlet, kg·m^{-3}.
μ = gas viscosity, Pa·s.
d_m = diameter of gas molecules, nm.
p_L = Langmuir pressure, Pa.
dp/dl = pressure gradient, Pa·m^{-1}.
M = molecular weight, kg·mol^{-1}.
R = universal gas constant, =8.314 J·mol^{-1}·K^{-1}.
k_B = Boltzmann constant, =1.38 × 10^{-23} J·K^{-1}.
T = ambient temperature, K.

ρ_s = density of shale matrix, kg·m^{-3}.

V_{std} = molar volume of gas under standard conditions, m^3·mol^{-1}.

q_L = Langmuir volume, m^3·kg^{-1}.

Φ_{co} = porosity of a core sample, dimensionless.

The expression of Fick diffusion (2) is referred to as bulk diffusion and represented by N_b.

The case study in literature [42] shows that although the equations of viscous flow and diffusion already contain variables varying with temperature, pressure, and other factors, they make sense within only a certain range of flow regimes and deviate from the actual situation within other range that is not taken into account. Introducing coupling coefficients to different flow mechanisms can help modify the correspondence between the mathematical models (i.e., those of viscous flow and diffusion) and Knudsen number and establish generalized models without segment processing as Kn varies.

In contrast to the coupling coefficients reported in published literatures [29, 31, 43, 44], the derivation of new coupling coefficients corresponding to the proposed flow mechanism scheme is performed, and the coupling coefficient of one certain flow mechanism will not be optionally set as 100%. The coupling coefficients of viscous flow, bulk diffusion, Knudsen diffusion, and surface diffusion are represented by $f_1(Kn), f_2(Kn), f_3(Kn),$ and $f_4(Kn)$ respectively, which are the functions of Kn. The expressions of the coupling coefficients are set according to the characteristics of flow regimes, where the following assumptions are used:

1. Let $f_1(Kn) + f_2(Kn) = 1/(1 + Kn)$ and $f_3(Kn) + f_4(Kn) = Kn/(1 + Kn)$ based on the molecular collision theory that the ratio of collision frequency between molecules to total collision frequency and that of molecule-wall collision frequency to total collision frequency are $1/(1 + Kn)$ and $Kn/(1 + Kn)$, respectively [30].

2. When Kn equals to 0, only viscous flow is assumed to exist [45], i.e., $f_1(Kn) = 1$ and $f_2(Kn) = f_3(Kn) = f_4(Kn) = 0$.

3. It is transition flow when $10^{-1} < Kn < 10$, and several diffusion processes play roles at the same time ([31, 46]; thus, $f_1(Kn)$ is assumed to be negligible at the logarithmic median of this range [29, 43], i.e., $f_1(Kn)$ is close to 0 when Kn > 1.

4. As Kn approaches to 0 or is sufficiently large, $f_2(Kn)$ is close to 0.

5. $f_3(Kn)$ is small when Kn < 1 and increases significantly when Kn > 1, until close to 1 in the range of Kn > 10 [29, 43].

6. In the whole range of flow regimes, $f_1(Kn), f_2(Kn), f_3(Kn),$ and $f_4(Kn)$ should all be nonnegative and change smoothly with Kn to embody the gradual evolvement of the flow as the condition varies.

Based on the above narrations, it physically defines that $f_1(Kn) = e^{-\alpha Kn}$, $f_2(Kn) = 1/(1 + Kn) - e^{-\alpha Kn}, f_3(Kn) = e^{-\beta/Kn},$ and $f_4(Kn) = Kn/(1 + Kn) - e^{-\beta/Kn}$, where α and β are dimensionless constants determining the bump levels of the variation curves. α and β are set as 5 and 1.8 [42], respectively, to further realize the compliance of the coupling coefficients with the narrated flow regime characteristics.

Hence, the mass flow in nanopores can be expressed as:

$$N = e^{-5Kn}N_D + \left(\frac{1}{1+Kn} - e^{-5Kn}\right)N_b + e^{-1.8/Kn}N_K + \left(\frac{Kn}{1+Kn} - e^{-1.8/Kn}\right)N_s \quad (5)$$

where N = mass flow in a pipe, $kg \cdot s^{-1}$.

The variation curves of the four coupling coefficients and $f_1(Kn) \times N_D$, $f_2(Kn) \times N_b$, $f_3(Kn) \times N_K$, and $f_4(Kn) \times N_s$ with Kn are depicted in
Figures 3 and **4** [42].

The benefits of introducing the above coupling coefficients to viscous flow and diffusion are significant:

1. It is clear that because $f_1(Kn), f_2(Kn), f_3(Kn)$, and $f_4(Kn)$ are all nonnegative, the segment processing of mathematical models can be avoided, i.e., Eq. (5) can be uniformly used for the coupling calculation in the scope of $0 < Kn < \infty$, without the need to change the functional forms by reason of the limited applicability of coupling coefficients.

2. Eq. (5) bridges the gaps between different flow regimes, i.e., the jumps of flow rates at the critical points between different regimes have vanished. Furthermore, the mathematical models are further constrained by virtue of the molecular collision theory to better reflect the basic flow regime knowledge.

3. Taking the viewpoints of Refs. [30, 32] as examples for comparison with this work, it should be noted that slip flow refers to the enhanced flow, including the part of original viscous flow and the other part called slippage effect which is represented by the non-zero velocities of the near-wall molecules due to gas-wall interactions. Therefore, it is more suitable to regard the ratio of gas–gas collision frequency to total collision frequency as the total coupling coefficient of viscous flow and bulk diffusion rather than that of the slip flow [30, 32].

4. The same examples [30, 32] are used for comparison. It is continuum flow when Kn approximates to 0. However, the coupling coefficient of slip flow is 1 when Kn = 0 in papers [30, 32], implying slip flow dominates in continuum flow regime, which contradicts the flow regime knowledge. This issue has been solved in this chapter.

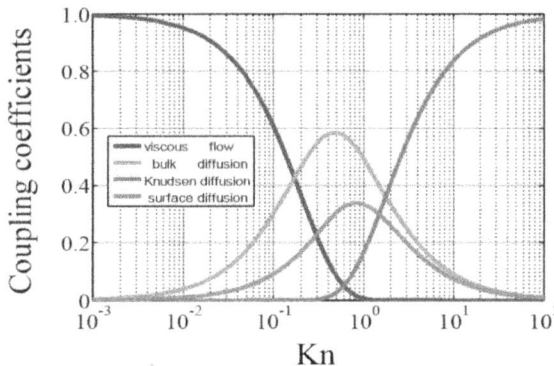

Figure 3.
Variation curves of the coupling coefficients (dimensionless) of viscous flow, bulk diffusion, Knudsen diffusion, and surface diffusion with Kn (dimensionless) [42].

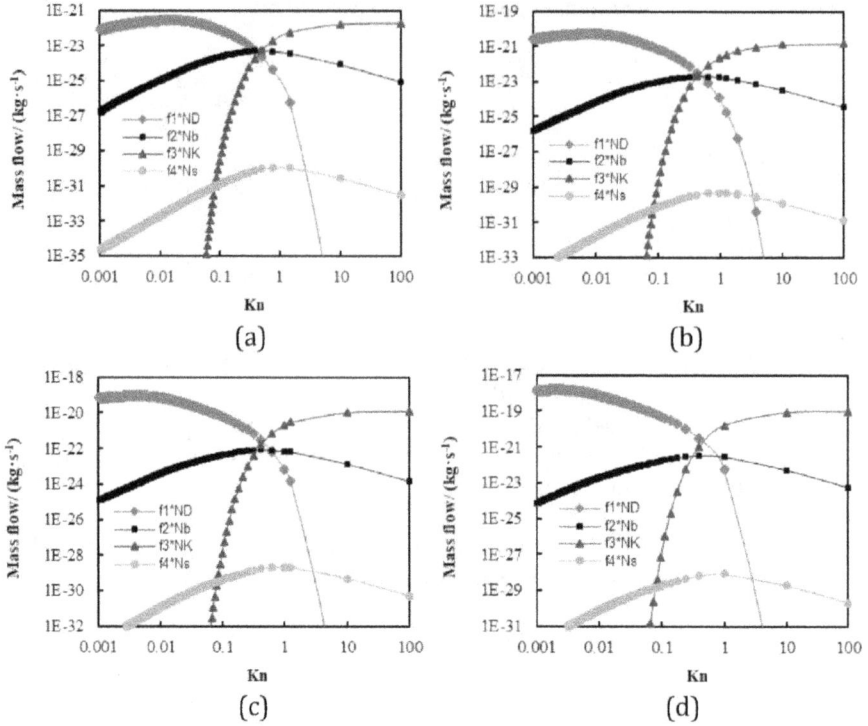

Figure 4.
Variations of viscous flow and diffusion with Kn (dimensionless) after introducing coupling coefficients for the gas flow in pores of (a) 5 nm, (b) 10 nm, (c) 20 nm, and (d) 40 nm at 353 K. $f_1{}^*N_D$, $f_2{}^*N_b$, $f_3{}^*N_K$, and $f_4{}^*N_s$ denote the results of viscous flow, bulk diffusion, Knudsen diffusion, and surface diffusion, respectively [42].

4. Coupled mathematical model in macroscopic-scale shale matrix

In this section, the experimental results of full-scale pore size distributions of real shale samples from a gas field are combined with the coupled flow model in nanopores to realize the upscaling transformation of the flow model into that in macroscopic-scale shale matrix by integration.

In the unitary model which is widely used for the flow estimation on a macroscopic scale [12, 18–22], indirect averaging methods are applied, e.g., the pore space of shale is assumed to be composed of a certain number of isodiametric pores, regardless of the pore size distributions. Some research [15, 47] used specific functions to characterize the probability density function of shale pore size distributions, with, however, assumed parameters for the purpose of conducting parameter sensitivity analysis. Here, the fitting parameters needed for the macroscopic form of the derived coupled flow model in nanopores are obtained by performing the experiments of pore size distributions of real shale samples from a gas field.

Michel et al. [15] and Xiong et al. [47] described the probability density function of shale pore size distributions as logarithmic normal distribution. Enlightened by their studies, the following expression is used to fit the experimental data of full-scale shale pore size distributions:

$$f(r_{in}) = \frac{1}{r_{in}\sigma\sqrt{2\pi}}e^{-0.5\left(\frac{\ln r_{in}-\eta}{\sigma}\right)^2}$$

(6)

Samples	η/dimensionless	σ/dimensionless
Ning 203-219	0.9428	1.0890
Ning 203-240	1.3530	1.2100
Ning 203-250	0.1207	0.4189
Average	0.8055	0.9060

Table 2.
Fitting results of η and σ.

where η = normal mean, dimensionless.

σ = variance, dimensionless.

Three kinds of experiments, i.e., the high-pressure mercury intrusion experiment, the liquid nitrogen adsorption experiment, and the low-temperature carbon dioxide adsorption experiment, were performed, and the full-scale pore size distribution data of the three shale samples from the Well "Ning 203", Longmaxi formation of Changning-Weiyuan district, Sichuan Basin of China, were obtained by stitching the three results together according to the effective range of each experiment, where the total volume of pores greater than 100 nm is attributed to the pore whose radius is closest to 100 nm in the experiments allowing for the difficulty of curve fitting caused by the severe fluctuations of the pore size data [42]. The values of η and σ are listed in **Table 2**. Because the samples "Ning 203-219", "Ning 203-240", and "Ning 203-250" are all taken from a depth interval of 2300-2400 m, the three groups of data in **Table 2** are averaged, i.e., η = 0.8055 and σ = 0.9060, to represent the typical shale pore size distribution in this depth range.

The number of single pipes in shale with the radius range of r_{in} to $r_{in} + dr_{in}$ is expressed in Eq. (7). By integrating in the entire pore size range, the flow rate in shale is obtained as Eq. (8):

$$\frac{10^{18}\phi_{co}V_{co}}{\pi r_{in}{}^2 L_{co}}f(r_{in})dr_{in} \tag{7}$$

$$Q = \int_{r_{in,\,min}}^{r_{in,\,max}} \frac{10^{18}N\phi_{co}V_{co}}{\pi r_{in}{}^2 L_{co}}f(r_{in})dr_{in} \tag{8}$$

where V_{co} = apparent volume of a core sample, m^3.

L_{co} = length of a core sample, m.

$r_{in,\,min}$ = lower limit of integration, which should be larger than 0.19 nm because the diameter of methane molecules is 0.38 nm [48].

$r_{in,\,max}$ = upper limit of integration.

The macroscopic-scale mathematical model of shale gas flow can be obtained by substituting Eqs. (5) and (6) into Eq. (8) as:

$$Q = \int_{r_{in,\,min}}^{r_{in,\,max}} \frac{10^{18}\left[e^{-5Kn}N_D + \left(\frac{1}{1+Kn} - e^{-5Kn}\right)N_b + e^{-1.8/Kn}N_K + \left(\frac{Kn}{1+Kn} - e^{-1.8/Kn}\right)N_s\right]\phi_{co}V_{co}}{\pi r_{in}{}^3 L_{co}\sigma\sqrt{2\pi}} e^{-0.5\left(\frac{lnr_{in}-\eta}{\sigma}\right)^2} dr_{in}$$

$$\tag{9}$$

Literature survey shows that there are several main upscaling methods of flow models from microscopic to macroscopic scale, i.e.:

Method (1): the commonly used unitary model [12, 18–22] as already mentioned.

Method (2): the sum method of calculating the permeability of every straight capillary tube [27].

Method	Description/equation	Advantages	Shortcomings
Unitary pipe model [12]	$q = nq_h = \frac{\phi A_b}{\pi R_h^2} q_h$ (q, total flow, $m^3 \cdot s^{-1}$; n, the number of pipes of hydraulic radius R_h, dimensionless; q_h, flow rate of single pipe, $m^3 \cdot s^{-1}$; ϕ, porosity, dimensionless; A_b, bulk surface area of porous media normal to flow direction, m^2; R_h, hydraulic radius, m)	Simple in formula and easy for calculation	Negligence of pore structure, e.g., different pore shapes, pore connectivity, etc.
Integral pipe model (this chapter)	$Q = \int\limits_{r_{in,\,min}}^{r_{in,\,max}} \frac{10^{18} N \phi_{co} V_{co}}{\pi r_{in}^3 L_{co} \sigma \sqrt{2\pi}}$ $\cdot e^{-0.5\left(\frac{\ln r_{in}-\eta}{\sigma}\right)^2} dr_{in}$ (Q, total flow, $kg \cdot s^{-1}$; $r_{in,\,max}$ and $r_{in,\,min}$, the minimum and maximum values of the inner radius of pipes, nm; N, mass flow in a pipe, $kg \cdot s^{-1}$; Φ_{co}, porosity of a core sample, dimensionless; V_{co}, apparent volume of a core sample, m^3; r_{in}, inner radius of a pipe, nm; L_{co}, length of a core sample, m; σ, variance, dimensionless; η, normal mean, dimensionless)	Make the consideration of various pore sizes happen; easy for calculation	Negligence of pore structure, e.g., different pore shapes, pore connectivity, etc.
Total addition model [27]	$q = \sum_i q_i$ (q, total flow, $m^3 \cdot s^{-1}$; q_i, flow rate of the ith single pipe, $m^3 \cdot s^{-1}$)	Consider the flow rate in every single pipe	Impractical to implement; negligence of pore structure, e.g., different pore shapes, pore connectivity, etc.
Model of statistical sum of permeability from each shape type [49, 50]	$(k_{app,pm})_{eff} = \left[\left(k_{app,pm}^{\frac{x}{100}}\right)_{slit} \times \left(k_{app,pm}^{1-\frac{x}{100}}\right)_{tube}\right]$ ($k_{app,pm}$, apparent permeability modified for ultratight porous media, m^2; $x/100$, percentage of rectangular slits pores; $(1-x/100)$, percentage of cylindrical pores)	Pore shapes, i.e., rectangular slits and cylindrical tubes, are taken into account	The quantification of the percentages of different pore types using image analysis tools is hard to implement; negligence of various pore sizes
3D fractal model [51]	Please refer to Eqs. (24)– (27) in literature [51] for the specific expressions where the formulas are complex	Multi-scale pore size distribution and tortuous flow line in 3D space of shale matrix are characterized	Many parameters to be determined; negligence of different pore shapes
Homogenization model [52, 53]	The homogenization method is used to upscale gas flow through two distinct constituents, a mineral matrix and organic matter. A gas flow in a two-constituent composite	The constituents, i.e., mineral matrix and organic matter, in shale are taken into account	Multiple assumptions; redundant processing for model establishment and solution

Method	Description/equation	Advantages	Shortcomings
	porous medium is considered, in which a microscopic unit cell is periodically repeated		
Pore network model [54, 55]	Generate pore network models by extracting pore structure information from real images or generate porous media by simulating the sedimentation and diagenesis processes and then incorporate relevant flow mechanisms into the gas flow models	Pore size distribution, anisotropy and low connectivity of the pore structure, etc. can be taken into account	Substantial work for model establishment; representativeness and verisimilitude of pore network models to the real pore structures remain a challenge

Table 3.
Comparison of upscaling methods from microscopic to macroscopic scale.

Method (3): the statistical sum method of the individual permeability from each shape type [49, 50].

Method (4): the 3D fractal model with variable pore sizes [51].

Method (5): the homogenization method to upscale gas flow through two distinct constituents, a mineral matrix and organic matter [52, 53].

Method (6): the pore network model including pore size distribution, anisotropy, and low connectivity of the pore structure, etc. in shale [54, 55].

The comparison among them is summarized in **Table 3**.

After reviewing the upscaling methods in **Table 3**, it is obvious that the method used in this work is not a bad compromise when compared to method (1) which is too simple and coarse, methods (2) and (3) where it is impractical and daunting to count the size/shape of every single pore with huge computational efforts, method (5) where complex processing for the model establishment and solution is needed, and methods (4) and (6) where redundant parameters/information about pore structure need to be assumed or obtained from multiple ways. Therefore, on the one hand, only the pore size distribution experiment is needed for the determination of the upscaling parameters in this chapter to make the consideration of various pore sizes happen. On the other hand, the derived model in this chapter is practical to operate, and the results can thus be readily obtained. However, it does not necessarily mean that there is no drawback for the upscaling method used. For example, although SEM images of the shale samples show that the pores in the organic matter are mostly circular [56], various types of pore shapes, e.g., cylindrical, triangular, rectangular shaped, etc., can be detected in shale samples [50, 57]. Singh et al. [50] concluded that the geometry of pores significantly influences apparent permeability of shale and diffusive flux. The study of effective liquid permeability in a shale system by Afsharpoor and Javadpour [58] confirmed that the assumption of simplified circular pore causes apparent permeability to be significantly overestimated and the discrepancy between the realistic multi-geometry pore model and the simplified circular pore model becomes more pronounced when pore sizes reduce and liquid slip on the inner pore wall is taken into account. Xu et al. [59] developed a model for gas transport in tapered noncircular nanopores of shale rocks and found the following: (1) pore proximity induces faster gas transport, and omitting pore proximity leads to the enlargement of the adsorbed gas-dominated region; (2) increasing taper ratio (ratio of inlet size to outlet size) and aspect ratio weakens real gas effect and lowers free gas transport; (3) moreover, it lowers the total transport capacity of the nanopore, and the tapered circular nanopore owns the greatest transport capacity, followed by tapered square, elliptical, and rectangular nanopores. To conclude, there is still much room for improvement of the upscaling method in this work in multiple aspects in future research.

5. Translation of lab-scale results into field-scale ones

With the properties of multi-scale pore structures and various reservoir modes, the shale gas reservoir is complex in reservoir space and occurrence modes, which in turn leads to different flow mechanisms in multi-scale spaces. Therefore, adopting single-scale equations and flow simulation methods will not accurately reveal the flow mechanism in complex shale gas reservoirs [60]. Jiao et al. [61] established an effective conversion relation between physical simulation parameters and field parameters based on similarity criterion to better simulate gas reservoir development. The ideas in literature [61] are narrated as follows.

First, considering the flow mechanism of shale gas in the reservoir, the selected characteristic physical parameters are permeability K, porosity ϕ, pore radius r, length L, original pressure p_i, flow rate of gas production q, gas compression factor Z, reservoir temperature T, standard temperature T_{sc}, and standard atmospheric pressure p_{sc}. According to the π theory, there are four basic dimensions named length dimension [L], mass dimension [M], time dimension [T], and temperature dimension [K]. Therefore, each of π is obtained, and field parameters are analyzed to deduce physical simulation parameters in the experiment according to the similarity criterion, as shown in **Table 4**.

Second, based on the similarity criterion, the conversion relation between physical simulation parameters and field parameters can be established, which is expressed as:

$$q_g = \frac{\pi r^2 K_{rg} K T_{sc} p_i^2}{LuZTp_{sc}} \left(\frac{LuZTp_{sc}}{\pi r^2 K_{rg} K T_{sc} p_i^2} q \right)_m \tag{10}$$

where m indicates that the parameters inside the brackets are for the physical simulation.

Finally, choose the core sample "Ning 211-1" for an example to conduct dynamic physical experiment under different conditions, which is used to verify the

Number	Similarity criterion	Similar attributes	Physical significance	Value of physical simulation	Actual value of reservoir
1	$\pi_1 = \phi$	Porosity similarity	Determine porosity	0.02–0.2	0.02–0.2
2	$\pi_2 = Z$	Compression similarity	Determine model gas	0.9–1.2	0.9–1.2
3	$\pi_3 = T/T_{SC}$	Temperature similarity	Determine model temperature	1–1.1	1.1–1.3
4	$\pi_4 = r/L$	Geometric similarity	Determine model size	0.3–1	0.3–1
5	$\pi_5 = p_{sc}/p_i$	Dynamic similarity	Determine original pressure of model	0.002–0.01	0.002–0.005
6	$\pi_6 = p_w/p_i$	Dynamic similarity	Determine conversion relation for bottom hole pressure	0–1.0	0.1–1.0
7	$\pi_7 = \frac{qLuZTp_{sc}}{\pi r^2 KK_{rg}T_{sc}p_i^2}$	Movement similarity	Determine production rate	0–0.5	0.1–0.3

Table 4.
Similarity criterion numerals of the gas reservoir physical simulation.

T_{sc}/K				293.15			
Φ_{co}				5.6%			
r_m/m				0.0127			
r/m				40			
L_m/m				0.0557			
L/m				20			
T_m/K				298.15			
T/K				353.15			
$p_i/(10^6 Pa)$	3.0745	4.0995	5.0800	6.5750	7.6500	10.2300	12.5900
$u/(10^{-5} Pa \cdot s)$	1.1560	1.1785	1.2030	1.2461	1.2817	1.3830	1.4944
Z_m	0.9481	0.9316	0.9163	0.8942	0.8795	0.8493	0.8294
Z	0.9747	0.9670	0.9602	0.9507	0.9445	0.9326	0.9254
$q_m/(ml/s)$	0.0344	0.0466	0.0570	0.0746	0.0877	0.1205	0.1450
$q_g/(ml/s)$	785.5063	1055.4281	1278.7645	1649.5661	1919.8761	2579.8383	3055.2185
$q/(ml/s)$	748.2798	1021.0548	1255.2453	1601.7201	1902.6402	2529.7590	3038.9881

Table 5.
Parameters for application.

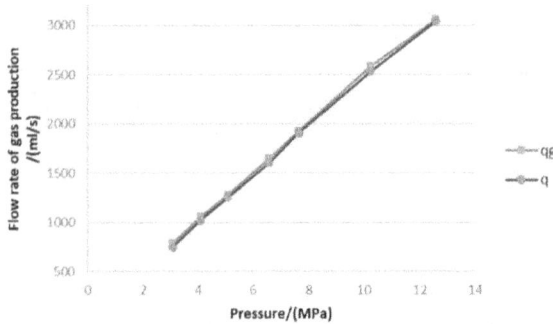

Figure 5.
Comparison of actual values of reservoir and predicted field results based on similarity conversion.

rationality of the similarity criterion. The related parameters, values of physical simulation (q_m), converted values of field (q_g), and actual values of reservoir (q) are presented in **Table 5**.

Figure 5 displays the curves of actual values of reservoir and predicted field results based on similarity conversion, the latter of which are calculated from the physical experiment. The results calculated by similarity criterion are basically consistent with the on-site tested data. It is expected that applying the similarity translation from physical simulation of gas reservoirs is capable of predicting the development performance effectively, showing the rationality of the translation method.

6. Conclusions

Based on our study in this chapter, the following conclusions have been reached:

1. A new concept "wall-associated diffusion" was introduced to the study of gas flow in shale nanopores, which has practical significance and multiple research

significance. By virtue of this concept, viscous flow, bulk diffusion, surface diffusion, and Knudsen diffusion were considered in the proposed flow mechanism scheme for nanoscale shale gas flow, with both division of mechanical mechanisms in nanopores and partition of flow space taken into account. Viscous flow and bulk diffusion belong to the bulk phase flow, which result from gas-gas interactions. In addition, surface diffusion and Knudsen diffusion are of boundary layer flow, which are associated with gas-wall interactions.

2. An easy-to-operate coupling method of the flow mechanism scheme containing four coupling coefficients and thus a coupled shale gas flow model in nanopores, which applies within the scope of full flow regimes and avoids segment processing, was proposed.

3. Based on the experimental data of pore size distributions of real shale samples from a gas field, a new coupled upscaling flow model in macroscopic-scale shale matrix with the experimentally determined fitting parameters was established. The model uses smooth functions to fit the full-scale pore size distribution results to facilitate the upscaling transformation of the model in nanopores into that in the macroscopic matrix.

4. A case study was presented to show how the lab-scale results are translated into field-scale ones, revealing the rationality of the translation method used.

In summary, sounder in theoretical bases and better in application effects, the proposed model is expected to be of practical significance for evaluating the gas flow capacity in shale matrix and guiding gas reservoir development in gas fields.

Acknowledgements

This work was supported by the National Science and Technology Major Project of the Ministry of Science and Technology of China (grant number 2017ZX05037 – 001); the Demonstration Project of the National Science and Technology Major Project of the Ministry of Science and Technology of China (grant number 2016ZX05062 – 002 – 001); and the Science and Technology Major Project of PetroChina (grant number 2016E–0611).

Author details

Zhiming Hu[1], Yaxiong Li[2*] and Yanran Li[1,3,4]

1 Research Institute of Petroleum Exploration and Development, China National Petroleum Corporation, Langfang, People's Republic of China

2 SINOPEC Petroleum Exploration and Production Research Institute, Beijing, People's Republic of China

3 University of Chinese Academy of Sciences, Beijing, People's Republic of China

4 Institute of Porous Flow and Fluid Mechanics, Chinese Academy of Sciences, Langfang, People's Republic of China

*Address all correspondence to: liyaxiong.syky@sinopec.com

IntechOpen

References

[1] Zhang et al. Foundation of Shale Gas Reservoir Development. 1st Ed. Beijing: Petroleum Industry Press; 2014

[2] Klinkenberg LJ. The Permeability of Porous Media to Liquid and Gases. In: API 11th Mid Year Meeting, May, Tulsa, USA

[3] Javadpour F. Nanopores and apparent permeability of gas flow in mudrocks (shales and siltstone). Journal of Canadian Petroleum Technology. 2009;**48**(08):16-21. DOI: 10.2118/09-08-16-DA

[4] Haghshenas B, Clarkson C R, Chen S. Multi-porosity multi-permeability models for shale gas reservoirs. In: SPE Unconventional Resources Conference Canada. Society of Petroleum Engineers; 2013

[5] Wu K, Li X, Wang C, et al. A model for gas transport in microfractures of shale and tight gas reservoirs. AICHE Journal. 2015;**61**(6):2079-2088. DOI: 10.1002/aic.14791

[6] Sun F, Yao Y, Li G, et al. A slip-flow model for multi-component shale gas transport in organic nanopores. Arabian Journal of Geosciences. 2019;**12**:143. DOI: 10.1007/s12517-019-4303-6

[7] Veltzke T, Thöming J. An analytically predictive model for moderately rarefied gas flow. Journal of Fluid Mechanics. 2012;**698**:406-422. DOI: 10.1017/jfm.2012.98

[8] Li Y, Li X, Shi J, et al. A nano-pore scale gas flow model for shale gas reservoir. In: SPE Energy Resources Conference. Society of Petroleum Engineers; 2014

[9] Mi L, Jiang H, Li J. The impact of diffusion type on multiscale discrete fracture model numerical simulation for shale gas. Journal of Natural Gas Science and Engineering. 2014;**20**:74-81. DOI: 10.1016/j.jngse.2014.06.013

[10] Song W, Yao J, Li Y, et al. Apparent gas permeability in an organic-rich shale reservoir. Fuel. 2016;**181**:973-984. DOI: 10.1016/j.fuel.2016.05.011

[11] Beskok A, Karniadakis GE. Report: A model for flows in channels, pipes, and ducts at micro and nano scales. Microscale Thermophysical Engineering. 1999;**3**(1):43-77

[12] Civan F. Effective correlation of apparent gas permeability in tight porous media. Transport in Porous Media. 2010;**82**(2):375-384. DOI: 10.1007/s11242-009-9432-z

[13] Moridis GJ, Blasingame TA, Freeman CM. Analysis of mechanisms of flow in fractured tight-gas and shale-gas reservoirs. In: SPE Latin American and Caribbean Petroleum Engineering Conference. Society of Petroleum Engineers; 2010

[14] Freeman CM, Moridis GJ, Blasingame TA. A numerical study of microscale flow behavior in tight gas and shale gas reservoir systems. Transport in Porous Media. 2011;**90**(1): 253-268. DOI: 10.1007/s11242-011-9761-6

[15] Michel GG, Sigal RF, Civan F, et al. Parametric investigation of shale gas production considering nano-scale pore size distribution, formation factor, and non-Darcy flow mechanisms. Society of Petroleum Engineers. 2011;**147438**: 38-46. DOI: 10.2118/147438-MS

[16] Deng J, Zhu W, Ma Q. A new seepage model for shale gas reservoir and productivity analysis of fractured well. Fuel. 2014;**124**:232-240. DOI: 10.1016/j.fuel.2014.02.001

[17] Deng J, Zhu W, Qi Q, et al. Study on the steady and transient pressure

characteristics of shale gas reservoirs. Journal of Natural Gas Science and Engineering. 2015;**24**:210-216. DOI: 10.1016/j.jngse.2015.03.016

[18] Roy S, Raju R, Chuang HF, et al. Modeling gas flow through microchannels and nanopores. Journal of Applied Physics. 2003;**93**(8): 4870-4879. DOI: 10.1063/1.1559936

[19] Tang GH, Tao WQ, He YL. Gas slippage effect on microscale porous flow using the lattice Boltzmann method. Physical Review E. 2005;**72**(5): 056301. DOI: 10.1103/PhysRevE.72. 056301

[20] Javadpour F, Fisher D, Unsworth M. Nanoscale gas flow in shale gas sediments. Journal of Canadian Petroleum Technology. 2007;**46**(10): 55-61. DOI: 10.2118/07-10-06

[21] Swami V, Settari A. A pore scale gas flow model for shale gas reservoir. In: SPE Americas Unconventional Resources Conference. Society of Petroleum Engineers; 2012

[22] Ziarani AS, Aguilera R. Knudsen's permeability correction for tight porous media. Transport in Porous Media. 2012; **91**(1):239-260. DOI: 10.1007/ s11242-011-9842-6

[23] Veltzke T, Thöming J. An analytically predictive model for moderately rarefied gas flow. Journal of Fluid Mechanics. 2012;**698**:406-422. DOI: 10.1017/jfm.2012.98

[24] Wu L. A slip model for rarefied gas flows at arbitrary Knudsen number. Applied Physics Letters. 2008;**93**(25):253103. DOI: 10.1063/ 1.3052923

[25] Singh H, Javadpour F. Langmuir slip-Langmuir sorption permeability model of shale. Fuel. 2016;**164**:28-37. DOI: 10.1016/j.fuel.2015.09.073

[26] Zhang P, Hu L, Meegoda JN. Pore-scale simulation and sensitivity analysis of apparent gas permeability in shale matrix. Materials. 2017;**10**(2):104. DOI: 10.3390/ma10020104

[27] Kuila U, Prasad M, Kazemi H. Application of Knudsen flow in modeling gas–flow in shale reservoirs. In: 9th Biennial International Conference and Exposition on Petroleum Geophysics, Hyderabad, India; 2013

[28] Dongari N, Sharma A, Durst F. Pressure-driven diffusive gas flows in micro-channels: From the Knudsen to the continuum regimes. Microfluidics and Nanofluidics. 2009;**6**(5):679-692. DOI: 10.1007/s10404-008-0344-y

[29] Rahmanian M, Aguilera R, Kantzas A. A new unified diffusion–viscous-flow model based on pore-level studies of tight gas formations. SPE Journal. 2012;**18**(01): 38-49. DOI: 10.2118/149223-PA

[30] Wu K, Chen Z, Wang H, et al. A model for real gas transfer in nanopores of shale gas reservoirs. Society of Petroleum Engineers. 2015. DOI: 10.2118/174293-MS

[31] Geng L, Li G, Zitha P, et al. A diffusion–viscous flow model for simulating shale gas transport in nano-pores. Fuel. 2016;**181**:887-894. DOI: 10.1016/j.fuel.2016.05.036

[32] Sun F, Yao Y, Li G, et al. Transport behaviors of real gas mixture through nanopores of shale reservoir. Journal of Petroleum Science and Engineering. 2019;**177**:1134-1141. DOI: 10.1016/j. petrol.2018.12.058

[33] Klinkenberg LJ. The Permeability of Porous Media to Liquids and Gases. In: API 11th Mid Year Meeting. American Petroleum Institute; 1941

[34] Hongkui G, Shen Y, Yan S, et al. Slippage effect of shale gas flow in

nanoscale pores. Natural Gas Industry. 2014;**34**(7):46-54. DOI: 10.3787/j. issn.1000-0976.2014.07.008

[35] Zhu Y, Xiangui L, Tieshu L, et al. A study of slippage effect of gas percolation in low permeability gas pools. Natural Gas Industry. 2007;**27**(5): 44-47

[36] Daixun C. Gas slippage phenomenon and change of permeability when gas flows in tight porous media. Acta Mech. Sin. 2002; **34**(1):96-100

[37] KeLiu WU, XiangFang LI, ZhangXing CHEN. The mechanism and mathematical model for the adsorbed gas surface diffusion in nanopores of shale gas reservoirs. Science China Press. 2015;**45**(5):525-540. DOI: 10.1360/N092014-00263

[38] Baisheng N, Li Z, Ma W. Diffusion micro-mechanism of coal bed methane in coal pores. Coal Geology & Exploration. 2000;**28**(6):20-22

[39] Lidong M, Hanqiao J, Junjian L, Ye T. Mathematical characterization of permeability in shale reservoirs. Acta Petrolei Sinica|Acta Petrol Sin. 2014; **35**(5):928-934. DOI: 10.7623/ syxb201405013

[40] Li Y, Liu X, Hu Z, et al. A new method for the transport mechanism coupling of shale gas slippage and diffusion. Acta Physica Sinica. 2017;**66**: 114702. DOI: 10.7498/aps.66.114702

[41] Li Y, Hu Z, Duan X, et al. The general form of transport diffusivity of shale gas in organic-rich nano-slits—A molecular simulation study using darken approximation. Fuel. 2019;**249**: 457-471. DOI: 10.1016/j. fuel.2019.03.074

[42] Li Y, Liu X, Gao S, et al. A generalized model for gas flow prediction in shale matrix with deduced coupling coefficients and its macroscopic form based on real shale pore size distribution experiments. Journal of Petroleum Science and Engineering. 2019;**187**:106712. DOI: 10.1016/j.petrol.2019.106712

[43] Dongari et al. Pressure–driven diffusive gas flows in micro–channels: From the Knudsen to the continuum regimes. Microfluidics and Nanofluidics. 2009;**6**:679-692. DOI: 10.1007/s10404-008-0344-y

[44] Wu K, Chen Z, Li X, et al. Flow behavior of gas confined in nanoporous shale at high pressure: Real gas effect. Fuel. 2017;**205**:173-183. DOI: 10.1016/j. fuel.2017.05.055

[45] Niu et al. Second–order gas–permeability correlation of shale during slip flow. SPE Journal. 2014;**19**: 786-792. DOI: 10.2118/168226-PA

[46] Chen et al. Channel–width dependent pressure–driven flow characteristics of shale gas in nanopores. AIP Advances. 2017;**7**:045217. DOI: 10.1063/1.4982729

[47] Xiong X, Devegowda D, Villazon M, et al. A fully-coupled free and adsorptive phase transport model for shale gas reservoirs including non-Darcy flow effects. SPE annual technical conference and exhibition. Society of Petroleum Engineers; 2012

[48] Etminan SR, Javadpour F, Maini BB, et al. Measurement of gas storage processes in shale and of the molecular diffusion coefficient in kerogen. International Journal of Coal Geology. 2014;**123**:10-19. DOI: 10.1016/j. coal.2013.10.007

[49] Fenton L. The sum of log-normal probability distributions in scatter transmission systems. IRE Transactions on Communication Systems. 1960;**8**: 57-67

[50] Singh H, Javadpour F, Ettehadtavakkol A, Darabi H. Nonempirical apparent permeability of shale. SPE Reservoir Evaluation and Engineering. 2014;**17**:414-424. DOI: 10.2118/170243-PA

[51] Cai J, Lin D, Singh H, Wei W, Zhou S. Shale gas transport model in 3D fractal porous media with variable pore sizes. Marine and Petroleum Geology. 2018;**98**:437-447. DOI: 10.1016/j.marpetgeo.2018.08.040

[52] Auriault JL. Heterogeneous medium. Is an equivalent macroscopic description possible? International Journal of Engineering Science. 1991;**29**:785-795. DOI: 10.1016/0020-7225(91)90001-J

[53] Darabi H, Ettehad A, Javadpour F, Sepehrnoori K. Gas flow in ultra-tight shale strata. Journal of Fluid Mechanics. 2012;**710**:641-658. DOI: 10.1016/0020-7225(91)90001-J

[54] Zhang P, Hu L, Meegoda J. Pore-scale simulation and sensitivity analysis of apparent gas permeability in shale matrix. Materials. 2017;**10**:104. DOI: 10.3390/ma10020104

[55] Wang L, Wang S, Zhang R, Wang C. Review of multi-scale and multi-physical simulation technologies for shale and tight gas reservoirs. Journal of Natural Gas Science and Engineering. 2017;**37**:560-578. DOI: 10.1016/j.jngse.2016.11.051

[56] Javadpour F, McClure M, Naraghi ME. Slip-corrected liquid permeability and its effect on hydraulic fracturing and fluid loss in shale. Fuel. 2005;**160**:549-559. DOI: 10.1016/j.fuel.2015.08.017

[57] Song W, Yao J, Ma J, Li A, Li Y, Sun H. Grand canonical Monte Carlo simulations of pore structure influence on methane adsorption in micro-porous carbons with applications to coal and shale systems. Fuel. 2018;**215**:196-203. DOI: 10.1016/j.fuel.2017.11.016

[58] Afsharpoor A, Javadpour F. Liquid slip flow in a network of shale noncircular nanopores. Fuel. 2016;**180**:580-590. DOI: 10.1016/j.fuel.2016.04.078

[59] Xu J, Wu K, Yang S, Cao J, Chen Z, Pan Y, et al. Real gas transport in tapered noncircular nanopores of shale rocks. AICHE Journal. 2017;**63**:3224-3242. DOI: 10.1002/aic.15678

[60] Sun H, Yao J, Yalchin E. Upscaling of gas transport in shale matrix based on homogenization theory(in Chinese). Science China Physics, Mechanics & Astronomy. 2017;**47**:114612. DOI: 10.1360/SSPMA2016-00531

[61] Chunyan JIAO, Huaxun LIU, Pengfei LIU, et al. Similarity criterion of the physical simulating experiment for the development performances of low-permeability tight gas reservoirs. Petroleum Geology and Oilfield Development in Daqing. 2019;**38**(1):155-161

www.ingramcontent.com/pod-product-compliance
Lightning Source LLC
Chambersburg PA
CBHW081223190326
41458CB00016B/5664